Lecture Notes in Computer Science   10680

Commenced Publication in 1973
Founding and Former Series Editors:
Gerhard Goos, Juris Hartmanis, and Jan van Leeuwen

More information about this series at http://www.springer.com/series/8637

Abdelkader Hameurlain · Josef Küng
Roland Wagner · Sherif Sakr
Imran Razzak · Alshammari Riyad (Eds.)

Transactions on Large-Scale Data- and Knowledge-Centered Systems XXXV

 Springer

Editors-in-Chief

Abdelkader Hameurlain
IRIT, Paul Sabatier University
Toulouse
France

Roland Wagner
FAW, University of Linz
Linz
Austria

Josef Küng
FAW, University of Linz
Linz
Austria

Guest Editors

Sherif Sakr
King Saud bin Abdulaziz University
Riyadh
Saudi Arabia

Alshammari Riyad
King Saud bin Abdulaziz University
Riyadh
Saudi Arabia

Imran Razzak
King Saud bin Abdulaziz University
Riyadh
Saudi Arabia

ISSN 0302-9743   ISSN 1611-3349   (electronic)
Lecture Notes in Computer Science
ISSN 1869-1994   ISSN 2510-4942   (electronic)
Transactions on Large-Scale Data- and Knowledge-Centered Systems
ISBN 978-3-662-56120-1   ISBN 978-3-662-56121-8   (eBook)
https://doi.org/10.1007/978-3-662-56121-8

Library of Congress Control Number: 2017957859

This Springer imprint is published by Springer Nature
The registered company is Springer-Verlag GmbH, DE
The registered company address is: Heidelberger Platz 3, 14197 Berlin, Germany

Preface

This volume contains five fully revised selected regular papers, covering a wide range of very hot topics in the field of data quality, social-data artifacts, data privacy, predictive models, and e-health. These include data-quality methodology for the public sector, analysis of the interaction between business and social-data artifacts, protocols for secure querying in the cloud while preserving the privacy of the participants, and algorithms for the prediction of norovirus concentration in drinking water.

We would like to sincerely thank the editorial board and the external reviewers for thoroughly refereeing the submitted papers and ensuring the high quality of this volume.

Special thanks go to Gabriela Wagner for her high availability and her valuable work in the realization of this TLDKS volume.

September 2017

Abdelkader Hameurlain
Josef Küng
Sherif Sakr
Imran Razzak
Alshammari Riyad
Roland Wagner

Editorial Board

Contents

The Data Quality Framework for the Estonian Public Sector and Its Evaluation

Establishing a Systematic Process-Oriented Viewpoint on Cross-Organizational Data Quality

Jaak Tepandi[1]([✉]), Mihkel Lauk[2], Janar Linros[2], Priit Raspel[3], Gunnar Piho[1], Ingrid Pappel[1], and Dirk Draheim[1]

[1] Large-Scale Systems Group, Tallinn University of Technology,
Akadeemia Tee 15a, 12618 Tallinn, Estonia
{jaak.tepandi,gunnar.piho,ingrid.pappel,dirk.draheim}@ttu.ee
[2] PricewaterhouseCoopers, Pärnu Mnt 15, 10141 Tallinn, Estonia
{mihkel.lauk,janar.linros}@ee.pwc.com
[3] Republic of Estonia Information System Authority,
Pärnu Maantee 139a, Tallinn 15169, Estonia
priit.raspel@ria.ee

Abstract. In this paper, we describe a data quality framework for public sector organizations is proposed. It includes a data quality methodology for the public sector organizations (DQMP) and a complex of state-level activities for improving data quality in the public sector. The DQMP comprises a data quality model, a data quality maturity model, and a data quality management process. Based on this framework, two guidelines have been developed: a data quality handbook for public sector organizations and guidelines for improving data quality in the public sector. The results of the evaluation of the data quality handbook on three basic Estonian state registers highlight lessons learned and confirm the usefulness of the approach. The paper aims at researchers investigating data quality in various environments and practitioners involved in information systems management, development, and maintenance.

Keywords: Data quality · Data management · Data quality maturity model · PDCA cycle

1 Introduction

ICT services are becoming more data oriented. It is not possible to provide high-quality eGovernment services with poor data, so data quality is a decisive factor in providing good services for citizens. Correct data is a prerequisite for providing continuous operation of vital services and thus functioning of the critical information infrastructure. According to the NIS Directive, security of network and information systems means the ability of those systems to resist actions that compromise the authenticity, confidentiality or other quality characteristics of data [26].

© Springer-Verlag GmbH Germany 2017
A. Hameurlain et al. (Eds.): TLDKS XXXV, LNCS 10680, pp. 1–26, 2017.
https://doi.org/10.1007/978-3-662-56121-8_1

Particularly important is the quality of personal data. Article 6 of Directive 95/46/EC stipulates that personal data must be accurate and, where necessary, kept up to date [27]. Similar principles are stated in Article 4 of General Data Protection Regulation (GDPR) [28]. Data quality in public information systems is important in many areas. In health information systems, available, reliable, timely, and valid data is a prerequisite for the provision of high-quality health services at all levels of the health care system [1]. Still, public data is not always of good quality. One of the barriers to creating and using big data is that datasets lack good metadata and the data quality is poor. As an example, the US government's 2009 initiative, data.gov, to make its large amounts of data readily available to the public, has been hindered by the difficulty of ensuring that the data are in a usable format and of homogeneous quality [12]. In an OECD paper about using performance information in the budget process [10] the authors underline that although OECD countries have reported a number of benefits from the use of performance information, most governments are finding it difficult to provide decision makers with good quality information.

The consequences of poor data quality are decreased performance, as well as increased risks and costs [16]. Examples of low-quality information include more than £210 million of benefits being paid to the deceased over a period of three years, as well as six million citizens wrongly taxed over a period of two years, totaling £3.8 billion [25]. As another example, in 2014 the California auditor's office issued a report that looked at accounting records at the State Controller's Office to see whether it was accurately recording sick leave and vacation credits. The audit found 200,000 questionable hours of leave due to data entry errors, with a value of $6 million [5]. Similar problems occur in many countries. As an example, the National Audit Office of Estonia assessed whether the information systems in the area of government of the Ministry of the Environment support the reliable, economical and sustainable collection and use of environmental information. The audit revealed that the correctness and reliability of the environmental data have been guaranteed to a considerable extent. The exception was found to be the Environmental Register, which contains considerably more errors than the other registers, although it is the main register in the Ministry of the Environment and should be the most reliable [6].

These and many other cases have motivated this research. One of its starting points has also been a data quality study project [37], which has been intitiated to improve data quality in public sector. The current study gives an overview of previous research done in public sector data quality management and its specific requirements. Based on the analysis of the previous research and on the constraints for the public sector data quality management, the public sector data quality framework is developed. This framework comprises state-level activities towards data quality and the data quality methodology for the public sector organizations (DQMP). The DQMP comprises the data quality model, the data quality maturity model, and the data quality management process.

To implement and evaluate the data quality framework, two guidelines have been developed: a data quality handbook for public sector organizations and guidelines for improving data quality in public sector. The framework was piloted on three Estonian base registers. The lessons learned confirm the usefulness of the approach. As the data quality management is closely related to overall data governance, we also analyze issues of overall data governance and how the data quality management framework fits in. The results aim at researchers investigating data quality in various environments and practitioners involved in information systems management, development, and maintenance.

The structure of the paper is as follows. Section 2 proposes a broad overview of previous research, constraints related to the public sector data quality framework, as well as the high-level structure of the framework. In Sect. 3, the data quality framework is developed. In Sect. 4, results of evaluation of the framework and its supporting guidelines are highlighted. In Sect. 5 we analyze relationship of overall data governance and the data quality management framework. We discuss further related work also in Sect. 6 and the results in Sect. 7.

2 Prior Research, Constraints, and a High Level Model

In this section, we give an overview of approaches to data quality, analyze specific requirements for public sector data quality management, and draft a high-level model of the public sector data quality management framework.

2.1 Approaches to Data Quality Management

As an example from an authoritative standardization body, the ISO/IEC 25012 standard [34] defines the data quality model that categorizes quality attributes into fifteen characteristics considered by inherent and system dependent points of view. Inherent data quality refers to the degree to which quality characteristics of data have the intrinsic potential to satisfy stated and implied needs when data is used under specified conditions. System dependent data quality refers to the degree to which data quality is reached and preserved within a computer system when data is used under specified conditions. These fifteen quality characteristics are accuracy, completeness, consistency, credibility, currentness, accessibility, compliance, confidentiality, efficiency, precision, traceability, understandability, availability, portability, and recoverability. When developing a data quality framework, international standards such as [34] should be taken into account.

The W3C Data Quality Vocabulary [49] is work in process that proposes relevant quality dimensions and ideas for corresponding metrics, including statistics, availability, processability, accuracy, consistency, relevance, completeness, conformance, credibility, and timeliness. The goal is not to define a normative list of dimensions and metrics but rather to provide a set of examples. This is an early draft of the W3C Data Quality Vocabulary that does not imply endorsement by the W3C membership. W3C standards are also potential sources for the development of an authoritative state-wide framework.

Many other data quality approaches exist. The following materials can be considered as potential sources for data quality characteristics used in the data quality model. The USAID tips for performance monitoring & evaluation in conducting data quality assessments comprise five key data quality standards: validity, reliability, precision, integrity, and timeliness [45]. The Data quality framework for Bank of England introduces the following data quality dimensions: relevance, accuracy, timeliness and punctuality, accessibility and clarity, comparability, and coherence [4]. The Data Quality Assessment Framework of International Monetary Fund has five dimensions of data quality: assurances of integrity, methodological soundness, accuracy and reliability, serviceability, and accessibility [32]. To encourage the use of performance information OECD countries, it should be relevant, of high quality, credible, timely, and presented in a simple and integrated manner [10]. The report of the EURIM sub-group on the quality of information describes the following data quality characteristics: accuracy, validity, reliability, timeliness, relevance, completeness, provenance, presentation, clarity and coherence, evidence, fitness for purpose. The report states that the public sector needs to rebuild its skills to manage and to use information at all levels [25]. Several other data quality models are provided in other sources, e.g., [8].

Besides proposing the characteristics to be included in the data quality model, the following list of sources illustrates some best practices to enhance data quality. These practices can be introduced on the level of the overall ICT governance level, or on the level of data quality management. Usability of these practices depends on the specific applications. The best practices in data quality include defining data standards for agencies, including metrics for accuracy, completeness, timeliness, and consistency; ensuring data quality in its source and in key checkpoints of the data route through the databases and agencies; utilizing unique persistent keys identifying the entity under consideration and data related to it; and implementing thorough data maintenance strategies to cope with continuously changing data [16]. These practices can be useful for improving data quality throughout an information system lifecycle. The McKinsey Global Institute has estimated that if the US health care industry were to transform its use of big data for more efficiency and quality, the outcomes could account for $ 300 billion to $ 450 billion in reduced health-care spending annually [35]. Steps to increase use of big data involve building global data banks on critical issues, engaging citizens and citizen science, building a cadre of data curators and analysts, and promoting virtual experimentation platforms [12]. Due to the overall high volume of public sector data, these steps may be useful for specific public sector systems.

Data quality can be enhanced by adequate modeling of the area of concern using business archetypes and archetype patterns based methodology for modeling of business domains. Business archetype is an abstract model describing some common term that consistently and universally occur in business domains and in business software systems. Business archetype pattern is collaboration between business archetypes. The product, process, party, party relationship, inventory, order, quantity and rule archetype patterns are members of this archetype patterns framework [42]. These practices are useful during an information system development phase. Negative consequences of bad government data can be reduced by spending money to replace and update substandard technology systems, involving more data scientists and analysts in government, building system controls to prevent inputting errors, making sure that employees inputting data are trained, creating or improving data governance, setting up auditing programs for data quality, and considering the interests of the other agencies that can possibly use that information [5]. These activities should be considered for inclusion into a public sector data quality framework.

Recommendations of the audit of information systems in area of government of Ministry of the Environment [6] included analyzing what information is needed most and using this knowledge for developing information systems, avoiding data forwarding via several channels, better application of information security requirements (such as integrity), more comprehensive usage of automatic controls, and deciding as soon as possible the fate of the problematic Environmental Register. These recommendations can be generalized and included in the data quality framework.

2.2   Constraints for the Framework

In this section, we introduce specific requirements for the data quality framework in the public sector. A mental model and a high-level structure for managing data quality in public sector organizations are proposed in Sect. 2.3.

Integrated Approach. To efficiently enhance data quality in public sector, an integrated data quality framework is needed which takes into account specific features for the public sector data management and provides an integrated approach to data quality management. This approach should address at least the following topics.

- What is data quality? The concept of data quality, including the quality characteristics.
- How to improve data quality? The factors underlying data quality and activities to enhance quality by working with these factors.
- How to organize the activities for data quality improvement? The data quality improvement process in an organization.
- How can the state contribute the data quality? State-level support activities for data quality.

In Sect. 2.1 we gave some examples of specific approches to data quality management – useful models, recommendations, and best practices. However, to our best knowledge, no integrated data quality framework taking into account the public sector characteristics and involving all the components mentioned above have been developed so far.

Relationship to ICT Governance in Public Sector. A data quality framework is part of the overall ICT governance. In the public sector the main IT governance tasks - evaluate, direct, and monitor [33] – are performed with the aid of legal acts, statutes, procedures, contracts, procedures, agreements, and so on. Therefore, the data quality framework must be based on these sources and comply with them. The aim of the current paper is to provide a complete data quality framework, so in some cases, selected topics are included that could also be tackled within a wider governance framework. If these topics have already been considered, then this simplifies implementation of the data quality framework.

Quality Requirements in Legal Acts. The data quality requirements are mostly stated in legal acts. Usually, these requirements are expressed in broad terms and do not quantify requirements. For example, the statement "The objective of maintenance of cadastre (1) The objective of the maintenance of the cadastre is to register information in the cadastre reflecting the value of land, the natural status of land and the use of land, and to ensure the quality of such information ..." in §1^1 of [20] presents a general requirement that is not quantified. In contrast, in real large databases, data quality is not absolute - there is usually a percentage of incorrect records. So an absolute requirement is typically not satisfied and one needs to determine an acceptable level of incorrect data. This may result in some degree of conflict with the law. The requirements often do not quantify non-functional requirements either. This is often in contrast with private sector databases, where the data owners set specific non-absolute requirements to data quality. Therefore, the data quality framework for the public sector cannot prescribe general precise data quality measures, but rather a mechanism by which the data owners will themselves establish these measures.

Once-Only Principle. Another specific public sector requirement is the once-only principle in data collection. According to this principle, public administrations should ensure that citizens and businesses supply the same information only once to a public administration. This is one of the underlying principles in the "eGovernment Action Plan 2016–2020" [24] and is stipulated by the Estonian legislation. As the data owners must often use data from outside sources, the consequence of the once-only principle is that these data owners do not have direct control over data quality in their databases. There are three main levels of control for the owner.

- Data production under the direct management of the data owner can be directly affected within the data quality framework.
- Data production within the data owner's organization, but not under the direct management of the data owner may be partially affected within the data quality framework.
- Data production outside the data owner's organization usually cannot be affected within the data quality framework. Still, there may be indirect possibilities, such as communicating with the outside registers, initiating measures on the state level, regulating the frequency of data transfer, etc.

Both the once-only principle and its consequence that the data owner does not have full control over data quality are specific to the public sector data management. They have not been considered in the approaches to data quality management given in the previous Sect. 2.1, so the framework should take them into account.

Requirements to the Data Quality Model. An essential component of the data quality framework is the data quality model, which comprises various quality characteristics and sub characterics. As an example, the ISO/IEC 25012 standard [34] defines the data quality model that categorizes quality attributes into fifteen characteristics such as accuracy, completeness, consistency, credibility, etc. In Sect. 2.1, numerous sources for data quality model characteristics were referenced. To design the data quality model for the public sector organizations, specific features of these organizations have to be taken into account. Analysis of requirements for the data quality model for the public sector organizations [37] established that the model should confirm to the following.

- The characteristics should be important for the public sector organizations
- The data owner should be able to influence the characteristics
- Quality characteristics included in the model must cover the whole area of data quality as much as possible; still, the number of characteristics should be kept low to enable efficient usage of the model
- The quality characteristics should be unique (should not duplicate each other)
- The characteristics should be measurable
- In weighing different data quality models, priority should be given to models developed by authoritative bodies involving public sector organizations

Minimality and Sufficiency. Last but not least, an important prerequisite for the public sector quality framework is its minimality and sufficiency. As the data quality framework is intended to serve as practical guidelines for public sector organizations, all its components should be as simple as possible, but still covering all the subject area.

These requirements to the data quality model resulting form specific features of public sector data management are taken as a baseline for selecting the quality model for the public sector in the forthcoming sections.

2.3   High Level Structure of the Data Quality Framework

To develop a data quality framework, the following three components are needed. First, the concept of data quality must be understood and elaborated. According to ISO/IEC 25012, data quality is the "degree to which the characteristics of data satisfy stated and implied needs when used under specified conditions" [34]. To work with data quality, we need to provide a quality model - set of characteristics that are used to specify requirements for and to evaluate data quality. Second, in order to manage quality, we need to understand what influences it and how to enhance it. As an example, a research model for critical success factors for data quality in information systems is provided in [50]. To use this kind of models in enhancing data quality in public sector databases, it is practical to make it gradational - to develop a data quality maturity model. Third, to work towards higher quality, we need to present a set of activities for moving towards better data quality within an organization. This can be expressed as a data quality management process that should embrace all critical success factors for data quality.

The previous three components are used within one organization or information system and form a data quality methodology for the public sector organizations. To make this methodology work for the whole public sector, a separate set of state-level activities must be elaborated. In principle, the DQMP might be applied to a state information system viewed as one large organization. In practice, however, the state as an organization is significantly different from an individual public sector organization, and so are the corresponding information systems. For this reason, this fourth constituent is designed taking DQMP into account, but still as a separate unit. This is the last component of the data quality framework - a complex of state-level activities for improving data quality in public sector. In summary, the high-level model of the data quality framework is depicted in Fig. 1.

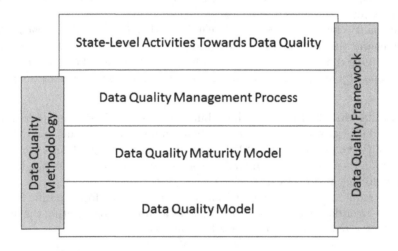

Fig. 1. High level model of the data quality framework for the public sector

3   The Data Quality Framework

In this section, we specify in more detail the main components of the data quality framework and present its summary view.

3.1   The Data Quality Model

Analysis of the data quality characteristics proposed in [4, 10, 12, 16, 32, 34, 37, 45, 49] shows that not all these characteristics satisfy the requirements for the data quality model provided above. As an example, the portability and recoverability characteristics from standard [34] relate to the system dependent technical issues, not to the properties of data as such. From the other side, not all the characteristics important for the public sector are present in these materials - for example, the once-only principle is very important for the public sector but seldom explicitly reflected as a data quality characteristics.

Taking into account all the constraints for the framework, the following nine characteristics of the data quality model for the public sector organizations were found to be the most relevant: accuracy, completeness, consistency, credibility, up-to-dateness, compliance, confidentiality, once-only principle, and non-redundancy. For a specific public sector organization, requirements or some other aspects for a selected characteristic may follow from an ICT governance framework, from security policies, from the data quality model, etc. This may simplify implementation of the data quality model in this organization. As there exists no uniform background with respect to pre-determined data quality characteristics and due to intention to keep the quality model complete, all the above characteristics are included in the current model. For each quality characteristic, the data quality model comprises its definition and explanations, main requirements, measures to be taken to achieve target level for this characteristic, as well as control questions for the responsible body. In what follows, we provide definitions for all quality characteristics and illustrate the other components of the model on the completeness example.

Definition of a Quality Characteristic. All definitions of quality characteristics of the model refer to a specific context of use. This means, inter alia, that the quality requirements are not absolute and usually there exists a cost-effective degree of required accuracy, completeness, consistency, and other characteristics, compare with Fig. 2. To support wider usage of the model, the definitions are based on ISO/IEC 25012 [34], whenever possible. To enhance understanding of the definition of a quality characteristic, two types of examples are provided: examples explaining the content of the characteristic, as well as examples illustrating relationships between characteristics. The latter examples demonstrate that the given quality model comprises characteristics which to a large degree are independent of each other.

The data accuracy characterizes the degree to which data attributes correctly represent the true value of the intended attributes of a concept or event in a specific context of use. It has two main aspects: syntactic accuracy and semantic accuracy. Syntactic accuracy requirement means that the data must confirm to the syntactic rules established in the domain under consideration. As an example, according to the Names Act of Estonia, a surname "B9rch" would be syntactically incorrect. Semantic accuracy is defined as the closeness of the data

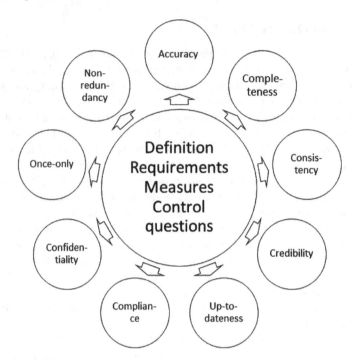

Fig. 2. The data quality model – characteristics and their components.

values to a set of values defined in a domain considered semantically correct. As an example, the data are semantically incorrect if a surname "Birch" is actually stored as "Pine".

The data completeness characterizes the degree to which data related to an entity have values for all attributes and instances required in a specific context of use. This does not imply that data for all attributes defined in a register or database must be provided. For instance, in a specific context of use, the first phone contact may be required to exist, but the second contact may be missing. Data completeness is independent of the other attributes - data may be complete and accurate, complete but inaccurate, etc.

The data consistency characterizes the degree to which data attributes are free from contradiction and are consistent with other data in a specific context of use. Consistency may refer both to data regarding one entity and to data regarding comparable entities. Data consistency is one of the prerequisites of information systems interoperability and it may be considered on syntactic, semantic, and other levels. As an example of consistency requirement, a person's date of graduating from a college cannot be earlier than the same person's date of birth. As another example, there is an inconsistency between the registers, if the same street is named 'Rävala street' in one register and 'Rävala road' in another register. In the latter case, inconsistency does not necessarily imply incorrectness, if both street names are actually in use.

The data credibility characterizes the degree to which data has attributes can be regarded as true and believable by users in a specific context of use. As an example, data provided by a governmental organization are usually considered credible. Data credibility or incredibility does not necessarily imply other quality attributes – for example, data may be accurate, complete, and consistent, but provided by an untrustworthy source.

The data up-to-dateness characterizes the degree to which data attributes are of the right age in a specific context of use. As an example, the Public Information Act of Estonia stipulates that the documents received by the agency and documents released by the agency shall be registered in a document register not later than on the working day following the day on which the documents are received or released. As with previous cases, different combinations of quality characteristics values are possible – for example, data may be inaccurate (some stored attributes of the registered documents have incorrect values), incomplete (some necessary attributes of the registered documents are missing), and up-to-date (registration was performed on the working day following the day on which the documents were received).

The data compliance characterizes adherence to current regulations, standards, contracts, conventions, and other rules relating to data quality in a specific context of use. As an example, the data in the Estonian environmental databases must confirm to the Environmental Register Act of Estonia. It is not possible to regulate everything by law. So defining the degree of data accuracy, consistency, and other quality characteristics sufficient for declaring data compliance remains the task of the chief processor of the environmental register.

The data confidentiality characterizes the degree to which data is accessible only by authorized users in a specific context of use. For instance, the Public Information Act of Estonia stipulates that public information is information which is obtained or created upon the performance of public duties provided by law or legislation issued on the basis thereof and that access to public information may be restricted pursuant to the procedure provided by law. Data may be confidential or non-confidential independently of their accuracy, completeness, consistency, etc.

The once-only principle states that information should be collected only once and then shared according to existing regulations and other rules. This reduces the administrative burden both for the users and businesses. As an example, many public sector information systems involve some personal or business data. The straightforward way to start development of such a system would be to collect, input, and maintain all data needed. It is much more complicated to build interfaces with central population and business registers, even assuming that these registers exist and their data is correct. In the long perspective, however, this straightforward development leads to multiple gathering of the same data, inaccuracies in the data due to scarce resources devoted to each data element, inconsistencies between databases, outdated data, etc. Thus, the once-only principle requirement supports to a certain extent other quality features such as data accuracy, consistency, up-to-dateness, etc. This relationship is however not absolute and it is possible for the data collected only once to be inaccurate, inconsistent, or outdated.

Non-redundancy characterizes the degree to which excess data is avoided in data structures and excess activities are avoided in data processes in a specific context of use. As an example, if data about each fishing permit of a fishing vessel in a fish resource dataset includes the vessel's detailed data, then this introduces redundancy which may lead to inconsistencies, obsolete data, or other problems. Non-redundancy requirement should be considered in the specific context of use – in some cases, redundancy in data or processing can be useful for emergency management, efficiency, or other reasons. Non-redundancy can relate to data structures or data processes of an individual database, or data and processes between different databases. While the once-only principle focuses on reducing the administrative burden for the citizens and businesses by eliminating the need for repeated collection of data, the non-redundancy requirement characterizes the extent of excess data and procedures in information systems. For example, the data about a fishing vessel may be collected only once, but still be used in a redundant way in the database. It is also possible to say that the once-only principle is a special case of non-redundancy, which due to its importance to the public sector data quality management is highlighted as a separate quality characteristic.

Requirements for the Data Quality Characteristics. The requirements for the data quality characteristics are given in laws and regulations, usually in broad terms without specifying quantitative values. This is in conflict with the real situation in registers, where for many valid reasons absolute data quality is rare. For data completeness, the main requirements stem from the Personal Data Protection Act of Estonia [21], from the Three-Level IT Baseline Security System (ISKE) [17], as well as from the laws and requirements for specific registers such as the Population Register Act [22]. As a typical example, the Population Register Act specifies attributes that must be present in the population register (such as the address data) and data that can be missing (such as the data about marital status). Still, the requirements concerning mandatory data are given in absolute terms, not setting any quantitative measures, such as allowed percentage of incomplete fields. In real records, accuracy and completeness of these data vary to a great extent [39].

Measures Needed to Achieve the Target Level for a Specific Characteristic. The measures needed to achieve target level for a specific characteristic comprise concretizing statutory requirements as needed, assessing compliance of data with respect to the requirements, and implementing measures to improve data quality if needed. To concretize statutory requirements, it may be necessary to set and agree on quantitative measures for the quality characteristics. These measures may stem from other regulations, contracts, service level agreements, and other sources. To assess compliance of data with respect to the requirements, evaluations must be performed according to requirements established in

Three-Level IT Baseline Security System or other relevant sources. These evaluations may be optional, periodic or real-time depending on the criticality of data. In case compliance is currently not ensured, all phases of data lifecycle should be analyzed to find possibilities for enhancing data quality. As an example, insufficient data completeness may be caused by inadequate data collection procedures, insufficient data controls, inadequate data updating processes, and other factors. Actions to be implemented will depend on the specific database organization, processes and technical resources, and they are different for each register. Measures for completeness assurance may comprise automatic completeness control procedures during data acquisition and update, periodic completeness checks and reporting, guides and training for personnel, user feedback, and so on.

Control Questions for the Responsible Body. Description of each characteristic comprises the control questions for the responsible body. The questions cover main requirements and measures to be taken to achieve target level for this characteristic. As an example, the control questions for data completeness address establishment of the completeness requirements based on regulations, processes, service level agreements, and other sources; ensuring data completeness in processing personal data; evaluating completeness of data with respect to requirements; taking measures to reduce incompleteness resulting from inadequate data collection procedures, insufficient quality controls, poor data processing procedures, etc.

3.2   The Data Quality Maturity Model

Prior experience of authors in public sector information systems and analysis of databases involved in the project [37] have made it clear that there is no one single factor affecting data quality. For example, an organization may have good software tools, but if the people are not motivated and management is not interested in supporting data quality, the quality remains low. There are many factors affecting data quality; a representative list is provided in [50]. The data quality maturity model is a tool for an information system owner to assess the maturity of the information system with respect to data quality and plan improvements. Based on analysis of factors influencing data quality [38,50] and taking into account the restrictions on the model provided above, the data quality maturity model has been based on the following five key factors: management and planning, organization and responsibilities, processes, knowledge and competencies, tools.

To enable management and evaluation of processes, structures, technology, etc., maturity models comprising a fixed number of maturity levels are used in many fields of activity (for example, [38,43]). A similar principle is proposed in the current framework. It involves five levels of maturity.

1. Reactive (initial). Data quality processes operate in unexpected ways, they are poorly controlled and tackled reactively, data quality requirements and the quality level are not known.

2. Controlled (repeatable). Need for data quality management has been acknowledged, data quality requirements have been established, repeatable procedures have been introduced.
3. Standardized (defined). Data quality processes are standardized, quality of the data is checked for compliance and confirms to requirements.
4. Managed. Sustainability of the processes has been achieved. Results of qualitative and quantitative data quality measurements are used for management of existing processes.
5. Optimized. Data quality processes are reviewed and regularly updated. Results of qualitative and quantitative data quality measurements are used for improving existing processes.

For each factor, several assertions are presented for evaluation of their truth values. Each assertion is associated with a certain maturity level. An overall structure of the data quality maturity model is given in Table 1.

Table 1. Structure of the data quality maturity model.

Key factor	Maturity level	Assertion
Management and planning	1	Assertion
		...
	2	Assertion
		...
	...	...
...		

To claim that an information system has a level L for a factor F, all the assertions for this factor belonging to all levels up to level L must be evaluated as true. All systems are assumed to be at least on level 1. An example in Table 2 presents assertions for the management and planning factor on levels 2, 3, and 4, together with their evaluations for two information systems: IS1 and IS2. The management and planning factor of IS1 is on level 3, as all assertions up to level 3 are evaluated to be true, and one evaluation on level 4 is evaluated as false. The management and planning factor of IS2 is on level 1, as all systems are assessed to be at least on level 1 and not all level 2 assertions are evaluated to be true.

The overall maturity level can be evaluated as the minimum of maturity levels for all factors. This gives an evaluation of maturity level within an individual system. To measure advancement for an information system or compare quality levels for different systems, an arithmetic mean or weighted average of levels for all factors may also be used. Table 3 depicts assessment of the overall maturity level for IS1 and IS2, based on hypothetical assessments of all key factors of the maturity model.

Table 2. Example assessment of a maturity level for one data quality factor.

Key factor	Level	Assertion	IS1	IS2
Management and planning	2	The information system data quality requirements are identified and documented	Yes	Yes
	2	All critical data subject to existing management policies are identified and documented	Yes	No
	3	Data quality is checked in terms of compliance	Yes	Yes
	3	Data quality policies are documented and published	Yes	Yes
	4	In the considered zone, data quality management is implemented on a uniform basis	Yes	Yes
	4	Data quality is measured and improved based on analysis of measurement results	Yes	Yes
	4	Data quality requirements are associated with higher level strategies or policies	No	Yes

Table 3. Example assessment of overall maturity level for two systems.

Key Factor	IS1	IS2
Management and Planning	3	1
Organization and responsibilities	2	3
Processes	3	4
Knowledge and competences	4	3
Tools	5	2
Overall maturity level rating	**2**	**1**

The assertions used in the data quality maturity model can be used for planning and implementing data quality improvements. As an example, organization IS2 from Table 2 might start identifying and documenting all critical data subject to existing management policies. The data quality maturity model is part of the overall ICT governance framework. It mostly elaborates and specifies data quality related issues of a governance framework, but more general topics are tackled as well when needed. As an example, the assertion "All critical data subject to existing management policies are identified and documented" from the above Table 2 may be considered when dealing with information security or when defining the overall ICT strategy. It is still included in the data quality maturity model because of its importance to data quality management. If the critical data are already identified and documented, for example as part of the ICT strategy development, then it is easy to identify this assertion as true.

3.3   The Data Quality Management Process

The data quality management process for the organizations is based on the OPDCA (Observe, Plan, Do, Check, Adjust/Act) Shewhart-Deming cycle [29].

Each step of this cycle involves one or more activities presented below. An organization implementing the data quality methodology for the public sector organizations should iteratively pass through all the steps.

Observe. At the beginning of each iteration, the existing situation is observed and the need for improvement is evaluated. The observation step includes three activities. During the first iteration, all these activities should be performed. In subsequent steps, some activities may be missed, such as re-evaluation of the maturity level if there have not been significant changes since the last iteration. The first observation activity is to define the scope of data which will be assessed and which quality will be improved. The next activity is zoning the organization in order to identify parts of the organization responsible for data quality in the previously defined scope. Finally, assessment of the existing level of data quality maturity is performed, to enable planning of improvement activities. This assessment provides estimations of the quality levels for each maturity factor and the overall data quality maturity level rating. Like the maturity model, the data quality management process is also related to the overall ICT governance framework. Many data quality process activities may already be performed according to overall data governance principles. As an example, zoning the organization with respect to data quality management may be a result of implementing wider ICT governance policies. Still, these policies are different in various public sector organizations, so that it is difficult to make any assumptions about their results. Zoning may also be implemented as a security related decision. In this case, zoning resulting from data quality management may be different from zoning related to security considerations. Having in view these uncertainties and with a goal to provide an integrated methodology, the zoning activity is thus included in the data quality process. If it has already been done, it is easy to use the results for the data quality process. An individual public sector organization is also welcome to shift some of the quality process activities into the overall ICT governance framework if it exists.

Plan. During the planning step, the target data quality maturity level is identified and the data quality improvement plan is developed. The target level of data quality maturity is based on the objectives of the organization for data quality within the defined data scope and on the assessed existing data quality level. For many public sector organizations, the overall optimum target is the standardized (defined) level. In a specific organization, this decision should be made as part of the overall ICT governance framework or as an activity of the data quality management process. It is not realistic to jump over two levels at once. If the overall target level is not yet attained, it is in most cases advisable to choose a target level that is one level higher than the existing one for the current iteration. The data quality improvement plan comprises activities needed to achieve the target level for the current iteration. For each maturity factor, the plan identifies assertions that were evaluated to be false, prescribes activities to make this assertion true, as well as evaluates resource needs and deadlines for these activities. In case some assertions are related to wider ICT governance issues, they may be resolved within these issues.

Do. The data quality improvement plan and its implementation may significantly vary depending on the organization, on the data, as well as on the existing and target quality maturity levels. While variations in organizations and data can be very individual, specific prescribed activities are given on each maturity level. A selection of such quality maturity level specific activities is presented below.

Every organization starts at least from the first, reactive (initial) level. There are no assumptions associated with it. For example, being on the first level may mean that the data quality requirements are not sufficiently understood and the activities related to data quality are not considered as such. This level does not necessarily imply bad data quality from the other stakeholders' viewpoint. On the second, controlled (repeatable) level the data quality requirements are understood and work on data quality is explicitly started. To improve from the initial to the controlled level, it is important to identify the data quality model used and determine quality requirements for each data quality characteristic. It is recommended to use all characteristics provided in the above data quality model; in case some of the characteristics are not needed, they may be omitted in justifying the reasons for exclusion. The requirements may arise from the law, contracts, service level agreements, etc. Critical data and respective data owners have to be determined. Data quality processes must be identified and documented. Usually, employees need data quality training. Work on data quality support tools must be initiated. On the third, standardized (defined) level, data quality is checked and confirms to the requirements. During movement to this level, data quality policies are documented, data owners are appointed, data quality related activities are carried out in data processing. Tools for improving data quality, e.g., consistency checking, are implemented. Achievement of the fourth, managed level assumes continuous data quality monitoring and maintenance. To advance to this level, data quality measurement procedures and tools, analysis of measurement results and measures taken in the case of deviation from the established requirements must be implemented. On the fifth, optimized level the key factors of the maturity model are continuously monitored and improved. Regular reviews and refinements of management and planning, organization and responsibilities, processes, knowledge and competencies, and tools are required to achieve this level.

Check. This step involves checking whether the data quality target levels for the current iteration and the overall final quality level have been achieved. To achieve the target level for the current iteration, all key factors of the data quality maturity model must confirm to this level.

Adjust (Act). The goal of the adjustment step is to fix the new status of the data quality and improve the data quality management process if needed.

The changes performed are acknowledged throughout the organization. The new data quality level is published in the Administration System for the State Information System RIHA [23] and set as the basis for the next iteration. In case the iteration target level was not achieved, the process is reviewed, the problems are identified and corrections are made.

3.4   State-Level Activities for Improving Data Quality

The real value of data quality in state registers is exposed in the cooperation of these registers. For example, in case of an integrated service that requests data from several databases, insufficient data quality in one of source registers may destroy the value of good quality data in other registers. Therefore it is not sufficient to aim at data quality only in some individual registers. The state needs data quality in most of its registers to provide better services to citizens, as well as information about the data quality before the development of integrated services. Still, most work directly related to data processing and its quality is done in the public sector organizations who own the registers, not on the state level. The state has several options to promote data quality in the registers.

First, it can initiate complementary legislative acts related to data quality. These initiatives depend on the situation in an individual country and are specific for different countries. For Estonia, the recommendations concerned obligation to provide a statute for a public sector registry, obligation to provide data quality requirements in the statutes, liabilities for the infringement of requirements provided in the statutes, and complementing the Administration System for the State Information System RIHA. Components related to data quality should be added to state-wide ICT strategies and policy documents. As the law establishment process follows its own complex rules, this option is not further specified in the current paper. Second, like owners of individual registers, the state needs to observe the existing level of data quality, plan the target level, implement the measures for achieving the target level, check the results, and improve the process if necessary. This leads to a state level OPDCA data quality improvement cycle. Examples of activities on each step of this cycle are presented below. Similarly to organizations, the state should iteratively pass through all the steps.

For Estonian state registers, it was recommended to input data about data quality management and the current data quality maturity level into the Administration System for the State Information System. Based on these data, it is possible to calculate the proportion of databases for which the data quality management has been applied, the proportion of organizations that have implemented data quality management procedures, and the average data quality maturity level.

The target maturity level should be realistic and cost-effective. Similar to the situation with individual registers, a too high level may result in excessive costs not justified by the results. An initial proposal is based on the logic of the maturity level - on the second level the requirements are understood, on the third level the system confirms to requirements. The next levels support continuity and improvement of data quality. As the initial goal for the state, the data quality should confirm to requirements, thus it is proposed that at least 60% of the state registers should reach the third maturity level by the year 2020. This preliminary target level will be concretized based on further experience of implementing the framework. When this level is achieved, it may be useful to work towards higher levels with some registers.

To achieve the target level, it is necessary to complement the Administration System for the State Information System RIHA with new features supporting registration of the quality levels for each Key Process Area of the maturity model and calculating the integrated metrics for all registers. A tool for the maturity level self-evaluation will be made available for register owners. The data quality handbook for state registers together with best practice examples and introductory tutorials will be made available. The competence model for the top managers will be complemented with the data quality related components. A subsequent study on the progress of data quality improvement in the country will help to evaluate the truthfulness of the quality related data in the Administration System. Results on data quality management will be available on the web. The data quality framework for the public sector organizations and the associated guidelines will be updated based on the lessons learned and feedback received.

As a summary, the data quality framework comprises a data quality methodology for the public sector organizations (DQMP) and a complex of state-level activities for improving data quality in public sector. The DQMP includes data quality maturity model, data quality model, and data quality management process, see Fig. 3.

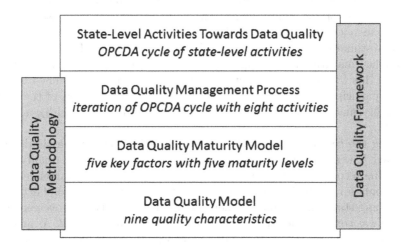

Fig. 3. The data quality framework for the public sector

4 Piloting and Evaluation of the Framework

To implement and evaluate the data quality framework, two guidelines have been developed: a data quality handbook for public sector organizations and guidelines for improving data quality in public sector. Using these guidelines, the framework was piloted on three Estonian registers: RIHA - the Administration System for the State Information System [23], the Population Register – one of the most important registers in Estonia [18], and the System of Address Details – one of the six support systems for the state information system [19].

4.1 Piloting Activities and Results

For each register, piloting included five activities of the first two steps – observing and planning – in the OPDCA cycle. Due to time restrictions of the study, the next steps of the OPDCA cycle were left for subsequent performing and review with the authors of the study. In what follows, the results of the activities are presented. The observing step of the data quality management process OPDCA cycle included three activities. The first activity was defining the scope of data. In all three cases, the whole register was selected for assessment and data quality improvement. The second activity of the observing step was zoning the organization. Zoning resulted in different cases, showing how variable the division of responsibilities can be in real organizations. For RIHA, the responsibility for data quality and mandate for changes were found to be located within the quality management system of the overall organization responsible for RIHA – the Information System Authority. For the Population Register, responsibility for data quality was within the department dealing with the register within the Ministry of the Interior. For the System of Address Details, the responsibility for data quality and mandate for changes were located within the responsible department of the Land Board. The third activity of the observing step was an assessment of the existing data quality maturity level. The assessment provided estimations of the quality levels for each maturity factor, the overall data quality maturity level rating, as well as the average score for maturity. Although this was not planned beforehand, the assessment determined that the registers participating in the study represented three basic levels of the overall data quality maturity level - low (maturity level 1), middle (level 3), and high (level 5).

During the second planning step of the OPDCA cycle, two activities were performed. The first activity was planning the target level of data quality maturity. As the recommended target is maturity level 3, the next target level was planned only for the register with the low maturity level. The second activity of the planning step was the development of the data quality improvement plan. The plan was composed for the register with the low maturity level. As an example, the plan included identifying and documenting the information system data quality requirements, identifying and documenting the critical data according to the targets set for the system, understanding and documenting the data quality processes, etc. During these activities, all organizations presented lists of data quality requirements from legislation and other sources, as well as gave answers to the control questions for the data quality model characteristics.

4.2   Lessons Learned

Piloting resulted in improving the framework, solving the participant problems, and giving answers to practical questions concerning the framework implementation. In what follows, these results are presented.

An important part of piloting was learning, understanding, and giving feedback on the framework, together with resultant improvements. Nearly 200 comments and suggestions made during piloting were used for restructuring the framework and improving its presentation. During the evaluation, the data quality management process was re-designed using the OPDCA cycle and the activities were re-scheduled. As an example, initially, it was believed that zoning should be in the end of the planning phase, perhaps due to an analogy with security management where this activity is located in middle of the security process. During piloting it became clear that for data quality management, zoning is a managerial decision and should be one of the starting activities. As discussed above, the best option would be to determine zoning from a wider data governance perspective, provided that such a perspective exists in the organization and zoning decisions resulting from different views (for example, strategy and security) are compatible.

Several problems were solved during the pilot implementations of the framework. One of these was that the participants expected precise numerical quality level values, for example acceptable percentages of incorrect data. In reality, a state-wide data quality model cannot give such overall numerical quality metrics, because they depend on specific registers and their requirements. Therefore it must be explicitly made very clear to the participants that the numerical quality level values have to be set by the organizations themselves and this activity is part of the data quality management process.

Another problem solved was that depending on the personalities of staff engaged in evaluation, the existing maturity level may easily be under- or overestimated. For example, although data quality procedures and documentation do not exist under explicit data quality headings, they may still be actually present under other titles. In this case, the data quality maturity level may actually be higher than the level initially perceived by the assessors. So the data quality team should search for all possible forms of data quality regulations.

An important practical question is how much resources needs to be devoted to the implementation of the framework. The number of persons active in piloting ranged from one to five. The pilot organizations measured time involved in the activities. It was found that the total time spent in piloting ranged from 18 h in the organization with the highest maturity level to 31 h in the organization with the lowest maturity level. Compared to presumable implementations in the future, this time may have been larger than usual as it included additional effort for giving feedback. At the same time, it could have been also somewhat smaller as the piloting organizations received support and help from the project group. Overall, it is foreseen that the magnitude of effort for the activities that were involved in piloting would remain similar in other organizations (about one person-week in total) and that this magnitude is acceptable.

5   The Data Quality Framework and Data Governance

In the Estonian state, there is an overall awareness for data governance that shows in diverse systematic data governance efforts. The delivered data quality framework is part and further fosters these overall data governance efforts and this section is about describing this relationship. Data governance is more than systematic data quality management [7,9,15], because data governance is driven by compliance issues.

The standard perspective of data governance is the perspective of the single organization. The perspective of the data governance framework of this article transcends this perspective, it is at the level of fostering and steering data governance in several bodies and organization involved in eGovernment. With respect to the single organization or enterprise, data governance always enacts a competitive edge to manage data as an asset, however, compliance adds a further impetus to data quality management. Therefore, data governance initiatives are organizational initiatives that have an impact on organizational processes. Ideally, there is a chain of governance initiatives in the organization. As a root, there should be a corporate governance initiative, compare with [40]. Then, there should be an IT governance initiative [33,48] that is aligned with corporate governance. Finally, the data governance initiative should be determined by the IT governance initiative. A natural way how such governance chain might emerge top down. A corporate governance initiative shows the need and triggers an IT governance initiative which then enacts a data governance initiative. If you enact a data governance initiative in its own right, its success will be crucially determined by the maturity of the already existing IT governance and corporate governance.

Today's implementations of governance initiatives are almost always truly process-oriented [11,30], i.e., process-orientation is not only a superficial device but a major crosscutting concern [13,14]. Consequently, process-orientation shows as a crucial element in the described data quality framework. We find it as OPCDA cycle in the data quality management process and, furthermore, as OPCDA cycle at the state-level activities. All that said, we can see the uniqueness of the described data governance framework. The data quality methodology, which encompasses three stages (data quality model, data quality maturity model, data quality management process), corresponds to the classical perspective of data governance, i.e., an organizational data initiative in service of overall corporate governance. But there is an encompassing state-level data governance initiative, which shows in the state-level activities of the data quality framework. The data quality methodology is designed towards and balanced against the state-level activities, which goes beyond classical data governance frameworks.

6   Related Work

An industrial-strength data governance framework is the IBM Data Governance Council Maturity Model [31]. The framework comes with a particular project-oriented attitude and breaks down the rational of data governance to concrete measurable values right from the outset, i.e., increasing revenue, lowering cost, reducing risks, increasing data confidence. The rational of this framework is clearly rooted in compliance and overall corporate governance, as can be seen in the effective elements of data governance it states, including organizational structures. In [36] Kathri and Brown develop a data governance framework that contains all classical elements of a governance framework. First, the rational is rooted in compliance, explicitly stating major legislative initiatives Sarbanes-Oxley Act and Basel II. Second, it is balanced against IT governance, very concrete against the key assets of the IT governance body of knowledge provided by Weill and Ross in [48]. Third, it identifies and elaborates different decision domains for data governance including data principles, data quality, meta data, data access and data lifecycle. The DGI Data Governance Framework [44] is an industrial-strength framework, which is an example of a full-fledged, process-oriented, corporate integrated data governance framework. In [47] Weber et al. are concerned with tailoring data governance according to the specific needs of organizations. They propose to exploit contingency theory to adjust data governance to specific parameters, both internal and external, compare also with [41]. In [46] Tountopoulos et al. address data governance in the presence of utilized cloud data services. They elaborate an accountability-based approach which targets stability concerning applicable legislation. Another recent work dealing with data governance in presence of cloud data services is Al-Ruithe et al. [2,3]. The work proposes concrete steps to establish data governance, where it maintains a balanced service consumer and service provider perspective.

7 Conclusion

Data quality is essential for making well-qualified decisions, for providing eGovernment services for citizens, and for the functioning of the critical information infrastructure. There are many reasons for insufficient data quality in public sector and absolute data quality is neither needed nor possible. Still, there is usually some optimum level of data quality that should be planned for, achieved and maintained. The proposed data quality framework addresses this need. It will be applied fully and needs to be tested with respect to applicability in public sector organizations outside Estonia. Besides its primary goal - improving data quality in public sector - it can be used as a component of a wider eGovernance framework and for evaluating the eGovernance level in surveys and indexes.

References

1. AbouZahr, C., Boerma, T.: Health information systems - the foundations of public health. Bull. World Health Organ. **83**(8), 578–583 (2005)
2. Al-Ruithe, M., Benkhelifa, E., Hameed, K.: Key dimensions for cloud data governance. In: Proceedings of FiCloud 2016 - the 4th International IEEE Conference on Future Internet of Things and Cloud. IEEE Press (2016)

3. Al-Ruithe, M., Benkhelifa, E., Hameed, K.: A conceptual framework for designing data governance for cloud computing. In: Shakshuki, E.M. (ed.) Proceedings of FNC 2016 - the 11th International Conference on Future Networks and Communications. Procedia Computer Science, vol. 94, pp. 160–167. Elsevier (2016)
4. Bank of England: Statistics and Regulatory Data Division. Data Quality Framework, March 2014. http://www.bankofengland.co.uk/statistics/Documents/about/dqf.pdf
5. Barrett, K., Greene, R.: The causes, costs and consequences of bad government data. In: GOVERNING, 24 June 2015
6. National Audit Office of Estonia: Maintenance and Development of Information Systems in Area of Government of Ministry of the Environment. Tallinn (2013)
7. Bordbar, B., Draheim, D., Horn, M., Schulz, I., Weber, G.: Integrated model-based software development, data access, and data migration. In: Briand, L., Williams, C. (eds.) MODELS 2005. LNCS, vol. 3713, pp. 382–396. Springer, Heidelberg (2005). https://doi.org/10.1007/11557432_28
8. Cai, L., Zhu, Y.: The challenges of data quality and data quality assessment in the big data era. Data Sci. J. **14**(2) (2015). http://datascience.codata.org/article/10.5334/dsj-2015-002/
9. Cong, G., Fan, W., Geerts, F., Jia, X., Ma, S.: Improving data quality - consistency and accuracy. In: Proceedings of VLDB 2007 - the 33rd International Conference on Very Large Data Bases. VLDB Endowment, pp. 315–326 (2007)
10. Curristine, T., Lonti, Z., Joumard, I.: Improving public sector efficiency - challenges and opportunities. OECD J. Budgeting **7**(1), 1–41 (2007)
11. Edwards Deming, W.: Out of the Crisis. MIT, Center for Advanced Educational Services (1982)
12. Desouza, K., Smith, K.: Big Data for Social Innovation. Stanford Social Innovation Review (2014)
13. Draheim, D.: The service-oriented metaphor deciphered. J. Comput. Sci. Eng. **4**, 253–275 (2010)
14. Draheim, D.: Smart business process management. In: 2011 BPM and Workflow Handbook, Digital Edition. Future Strategies, Workflow Management Coalition, pp. 207–223 (2012)
15. Draheim, D., Nathschläger, C.: A context-oriented synchronization approach. In: Electronic Proceedings of the 2nd International Workshop on Personalized Access, Profile Management, and Context Awareness: Databases (PersDB 2008) in Conjunction with the 34th VLDB Conference, pp. 20–27 (2008)
16. Dun & Bradstreet: Transparent Government Demands Robust Data Quality (2009). http://www.dnb.com/content/dam/english/dnb-solutions/sales-and-marketing/transparent_government_demands_data_quality.pdf
17. Estonian Information System Authority (RIA). Three-level IT Baseline Security System ISKE. https://www.ria.ee/en/iske-en.html
18. Estonian Parliamant. Population Register Act. https://www.riigiteataja.ee/en/eli/523032017001/consolide
19. Estonian Parliamant: System of Address Details (In Estonian). https://www.riigiteataja.ee/akt/113102015002
20. Estonian Parliament: Land Cadastre Act, Estonia (2016). https://www.riigiteataja.ee/en/eli/ee/Riigikogu/act/522062016005/consolide
21. Estonian Parliament: Personal Data Protection Act. Estonia, 16 January 2016. https://www.riigiteataja.ee/en/eli/ee/Riigikogu/act/507032016001/consolide
22. Estonian Parliament: Population Register Act, Estonia, 01 February 2016. https://www.riigiteataja.ee/en/eli/ee/Riigikogu/act/504022016005/consolide

23. Estonian State Information Management System Authority (RIHA). https://www.ria.ee/en/administration-system-of-the-state-information-system.html
24. European Commission: EU eGovernment Action Plan 2016–2020 - Accelerating the digital transformation of government (2016). http://ec.europa.eu/newsroom/dae/document.cfm?doc_id=15268
25. EURIM: Improving the Evidence Base - The Quality of Information Status Report and Recommendations of the EURIM Sub-group on the Quality of Information (2011)
26. European Parliament: DIRECTIVE (EU) 2016/1148 OF The European Parliament and of the Council of 6 July 2016 concerning measures for a high common level of security of network and information systems across the Union (2016)
27. European Parliament: Directive 95/46/EC of the European Parliament and of the Council on the protection of individuals with regard to the processing of personal data and on the free movement of such data, 24 October 1995
28. European Parliament: Regulation (EU) 2016/679 of the European Parliament and of the Council on the protection of natural persons with regard to the processing of personal data and on the free movement of such data, and repealing Directive 95/46/EC, 27 April 2016
29. Foresight University: Shewhart-Deming's Learning and Quality Cycle. http://www.foresightguide.com/chapter-8/shewhart-and-deming/
30. Hammer, M., Champy, J.: Reengineering the Corporation: A Manifesto for Business Revolution. HarperCollins Publishers, New York (1993)
31. IBM: The IBM Data Governance Council Maturity Model - Building a Roadmap for Effective Data Governance. IBM Corporation (2007)
32. International Monetary Fund: Data Quality Assessment Framework - Generic Framework (2012). http://dsbb.imf.org/images/pdfs/dqrs_Genframework.pdf
33. International Organization for Standardization: International Standard ISO/IEC 38500:2015. Information technology - Governance of IT for the Organisation, ISO (2015)
34. International Organization for Standardization: International Standard ISO/IEC 25012:2008. Software Engineering - Software Product Quality Requirements and Evaluation (SQuaRE) - Data quality model, ISO/IEC (2008)
35. Kayyali, B., Knott, D., Van Kuiken, S.: The Big-Data Revolution in US Health Care - Accelerating Value and Innovation, April 2013. http://www.mckinsey.com/industries/healthcare-systems-and-services/our-insights/the-big-data-revolution-in-us-health-care
36. Khatri, V., Brown, C.V.: Designing data governance. Commun. ACM 53(1), 148–152 (2010). ACM Press
37. Linros, J., Lauk, M., Tepandi, J., Kähari, V.: AS PricewaterhouseCoopers Advisors. Data Quality Study - Final Report (in Estonian). Information System Authority, 26 August 2016. https://www.ria.ee/public/publikatsioonid/Andmekvaliteedi_uuringu_lopparuanne.pdf
38. Loshin, D.: The Practitioner's Guide to Data Quality Improvement. Morgan Kaufmann, San Francisco (2010)
39. National Audit Office of Estonia: Population Data in National Registers (in Estonian). Tallinn (2002)
40. Organisation for Economic Co-operation and Development. G20/OECD Principles of Corporate Governance. OECD (2015)
41. Otto, B.: Data governance. Bus. Inf. Syst. Eng. 3, 241–244 (2011)
42. Piho, G., Tepandi, J.: Business domain modelling with business archetypes and archetype patterns. In: Information Modelling and Knowledge Bases, XXIV (2013)

43. Software Engineering Institute: CMMI for Development, version 1.3. http://cmmiinstitute.com/cmmi-models

44. Thomas, G.: The DGI Data Governance Framework. The Data Governance Institute (2016)

45. US Agency for International Development: Performance Monitoring & Evaluation Tips Conducting Data Quality Assessments, USAID (2010). http://pdf.usaid.gov/pdf_docs/Pnadw118.pdf

46. Tountopoulos, V., Felici, M., Pannetrat, A., Catteddu, D., Pearson, S.: Interoperability analysis of accountable data governance in the cloud. In: Cleary, F., Felici, M. (eds.) CSP 2014. CCIS, vol. 470, pp. 77–88. Springer, Cham (2014). https://doi.org/10.1007/978-3-319-12574-9_7

47. Weber, K., Otto, B., Osterle, H.: One size does not fit all - a contingency approach to data governance. ACM J. Data Inf. Q. 1(1), 4 (2009). ACM Press

48. Weill, P., Ross, J.W.: IT governance - How top Performers Manage IT Decision Rights for Superior Results. Harvard Business School Press, Boston (2004)

49. World Wide Web Consortium: Data Quality Vocabulary - W3C First Public Working Draft 25 June 2015. https://www.w3.org/TR/2015/WD-vocab-dqv-20150625/

50. Xu, H.: Factor analysis of critical success factors for data quality. In: Proceedings of AMCIS 2013 - the 19th Americas Conference on Information Systems (2014)

Bridging the Gap Between the Business and Social Worlds: A Data Artifact-Driven Approach

Zakaria Maamar[1(✉)], Vanilson Burégio[2], Mohamed Sellami[3],
Nelson Souto Rosa[4], Zhengshuai Peng[3], Zachariah Subin[3], Nayana Prakash[3],
Djamal Benslimane[5], and Roosevelt Silva[6]

[1] Zayed University, Dubai, UAE
zakaria.maamar@zu.ac.ae
[2] Federal Rural University of Pernambuco, Recife, Brazil
[3] ISEP Paris, Paris, France
[4] Federal University of Pernambuco, Recife, Brazil
[5] Claud Bernard Lyon 1 University, Lyon, France
[6] Recife Center for Advanced Studies and Systems, Recife, Brazil

Abstract. The widespread adoption of Web 2.0 applications has forced enterprises to rethink their ways of doing business. To support enterprises in their endeavors, this paper puts forward business-data artifact and social-data artifact to capture, respectively, the intrinsic characteristics of the business world (associated with business process management systems) and social world (associated with Web 2.0 applications), and, also, to make these two worlds work together. While the research community has extensively looked into business-data artifacts, there is a limited knowledge about/interest in social-data artifacts. This paper defines social-data artifact, analyzes the interactions between business- and social-data artifacts, and develops an architecture to support these interactions. For demonstration purposes, an implementation of a socially-flavored faculty-hiring scenario is discussed in the paper. The implementation calls for specialized components known as social machines that support artifact interaction.

Keywords: Business-data artifact · Social-data artifact · Social machine · Web 2.0 application

1 Introduction

Continuous progress of Information and Communication Technologies (ICT) has allowed enterprises to deploy better e-business applications. Among these ICTs, the Web now is a robust platform upon which cross-border business processes are deployed. *"A process is nothing more than the coding of a lesson learnt in the past, transformed into a standard by a group of experts and established as a mandatory flow for those who must effectively carry out the work"* [32].

© Springer-Verlag GmbH Germany 2017
A. Hameurlain et al. (Eds.): TLDKS XXXV, LNCS 10680, pp. 27–49, 2017.
https://doi.org/10.1007/978-3-662-56121-8_2

An online presence on the Web along with efficient business processes have become mandatory for any enterprise. To ensure that business processes perfectly capture an enterprise's best practices when satisfying and addressing end-users' requirements and concerns, respectively, IT practitioners use multiple techniques such as interviews and observations. Unfortunately, most techniques assume that end-users are familiar with the technical jargons of IT practitioners, which is not always the case. As a consequence, end-users (i.e., non-IT practitioners) do not properly express their requirements and concerns, which has a negative impact on the efficiency of IT practitioners' solutions. To mitigate this impact, business-data artifacts (for short, Business Artifacts (BA) are hailed for their particular role in bridging the gap between IT practitioners and non-IT practitioners [10,13,17]. BA is a concrete, identifiable, self-describing chunk of information that can be used by a business person to actually run a business [31]. BAs have life-cycles consisting of states and vary from one application domain to another.

A total reliance of enterprises on business processes might not be the sole response to today's challenges like globalization and capital scarcity [19]. Along with a smart management of business processes [12], Web 2.0 technologies (e.g., REST and JSON) and Web 2.0 applications (e.g., LinkedIn and Facebook) are, here, to help enterprises tap into social media's opportunities so they reach out to more customers, profile customers, gauge customers' satisfaction levels, etc. Enterprise 2.0 (or social enterprise [15]) is the one that successfully connects the business and social worlds together in a way that the specificities of both the business community (e.g., how to handle a new tax law) and the social community (e.g., how to respond to a customer's tweet) are accommodated. In [25], we introduce the concept of social-data artifact (for short Social Artifacts (SA)[1] that will work hand-in-hand with BAs. On the one hand, the business world includes BA-based processes that enterprises roll out in response to specific events (e.g., customer order) and in compliance with specific regulations (e.g., tax laws). On the other hand, the social world includes SAs that are created in response to events (e.g., social media campaign) and activities (e.g., product recommendation) that happen over Web 2.0 applications. To illustrate BA/SA collaboration, we adopt faculty hiring as a running example. The example consists of two parts: (*i*) business part is concerned with for instance, securing the necessary administrative approvals, defining criteria for shortlisting applicants, and preparing offers to applicants, and (*ii*) social part is concerned with for instance, posting job openings on Web 2.0 applications, answering applicants' questions, and screening applicants' social pages. How to identify the necessary SAs and how to ensure their interactions with BAs are key questions that we address in this paper.

In our previous work [25], we briefly defined SA, sketched an architecture for enterprise 2.0, briefly compared BA and SA, and implemented a proof-of-concept for creating SAs. In this paper, we revise the definition of SA using 2 perspectives referred to as data and operation, provide additional details on

---

[1] Some authors use the term of Web 2.0 artifact [3].

this architecture in terms of components' roles, define the messages that the business and social worlds exchange in the context of enterprise 2.0, establish correspondences between business and social artifacts, discuss how social artifacts are obtained, and finally, implement the faculty hiring running-example to illustrate these messages and analyze the data exchange. The remainder of this paper is organized as follows. Section 2 presents BAs in brief and then an overview of the use of social software in the context of enterprises. Section 3 details the BA-SA collaboration using a running example, presents the support architecture for this collaboration, and analyzes the interaction messages. A system implementing this example is detailed in Sect. 4. Finally, Sects. 5 and 6 list some future work initiatives and draws some conclusions, respectively.

2   Background

This section provides an overview of BAs and then, discusses the role of social software in supporting enterprise 2.0.

2.1   Business-Data Artifacts in a Nutshell

BA is a concrete, identifiable, self-describing chunk of information that can be used by a business person to actually run a business [31]. Several initiatives on BAs are reported in the literature [1,2,10,17,22,29–31,34,36,37].

Kumaran et al. [22] acknowledge the importance of identifying BAs in domain-specific applications. To this end, they assist IT practitioners identify the necessary BAs through guidelines. Narendra et al. [30] present an approach to model business processes using context-based artifacts and Web services. The authors abstract processes using models that are expressive (i.e., easy to grasp) for non-IT practitioners. These models could be based on BAs. As stated earlier, the technical jargon does not help non-IT practitioners properly express their needs. Maamar et al. [24] examine how to derive BAs from business requirements, so they present an approach for BA discovery and modeling. The approach uses a bottom-up analysis to determine fine-grained data, which are afterwards aggregated into clusters where each cluster being a potential BA. Next, the analysis derives the operations that act upon the discovered data clusters. Finally, the data and operation clusters are grouped into final BAs. Nigam and Caswell [31] define BAs along with their life-cycles in business modeling. In this case, a BA is a formal structure suitable for business people so they can manage, analyze, and control their daily business operations. Last but not least, Popova et al. [34] acknowledge the role of BAs in modeling business processes and propose a set of methods to discover BAs' life-cycles. The methods are implemented as software plug-ins for ProM (www.promtools.org), a generic open-source framework for supporting different process mining techniques.

In line with the aforementioned initiatives, we consider a BA as a chunk of meaningful information related to a certain business operation. Figure 1 illustrates the life-cycles of three BAs: order, customer, and bill. These BAs are used in

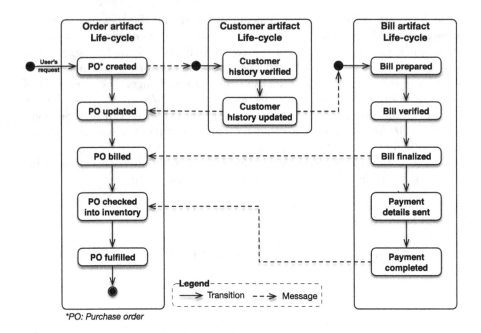

Fig. 1. Representation of a BA-based purchase-order BP (adapted from [30])

the context of a simplified version of the traditional purchase-order scenario [30] where different steps fulfilling a purchase order (e.g., creation/update by the customer and billing) are illustrated. In this figure, rounded rectangles, plain lines, and dashed lines represent artifact states, transitions within the same life-cycle, and messages between states in separate life-cycles, respectively.

2.2 Enterprise 2.0

In addition to regular Web 2.0 applications like Facebook and Instagram, enterprise 2.0 has the opportunities to capitalize on specific software referred to as social software. We discuss social software and then illustrate how it is adopted in two specific areas: business process management and coordination. According to Global Industry Analysts, Inc. "*The global expenditure on Enterprise Web 2.0 is forecast to reach $5.7 billion by 2015, driven by expanding broadband capabilities, decreasing prices, improving performance of networks, and the development of advanced, highly interactive Web 2.0 applications*" [40] and "*... the top 15 Web 2.0 vendors will spend $50 billion in 2015 on servers, networks, and other infrastructure, up from $38 billion in 2014 and $30 billion in 2013*" [39].

According to Warr [38], social software includes a large number of tools used for online communication, e.g., instant messaging, text chat, Internet fora, weblogs, Wikis, social network services, social guides, social bookmarking, social citations, social libraries, and virtual worlds. Erol et al. [14] mention that the roots of social software can be traced back to the 40s and, also, add that impressive results are obtained without a central plan or organization. Social software supports interactions between individuals who do not necessarily know each other and are not organized in any hierarchy. According to Bruno et al. [5], social software has four characteristics: weak ties, social production, egalitarianism, and mutual service provisioning.

A number of projects demonstrate the use of social software in business process management. The AGILe busIness PrOcess (AGILIPO) project embeds social software into business process tools [35]. Because business processes are inherently incomplete, human involvement is usually required. AGILIPO considers that stakeholders play three different roles, namely executor, modeler, and developer. Brambilla et al. [4] propose a specific notation to design social business processes. The notation includes new types of events and tasks like broadcast, post, and invite to activity. Grim-Yefsah et al. [18] acknowledge the existence of informal networks in the work environment and that people rely on these networks to complete their duties. The informal networks perfectly co-exist with regular networks based on formal relations (e.g., supervision). In fact, it is acceptable that the "formal" executor of a task informally seeks assistance from peers in the organization. Busch and Fettke [8] point out that business process management can be improved through social-network analysis in such a way that existing relations among employees are exposed. An accurate assessment of how a work is achieved is to analyze "as-is" processes (what is really happening) *versus* "to-be" processes (what has been defined).

Regarding the role of social software in improving coordination, multiple projects exist as well. Dengler et al. [11] promote social software like Wikis and networks to ensure the success of collaborative process activities. These activities take users' profiles posted on social networks so that appropriate users are identified. Dengler et al.'s approach reuses best practices to facilitate the creation of process models. Masli et al. [27] propose a social coordination mechanism based on the notion of "tentative event" in order to schedule meetings. Meetings are defined with attributes (e.g., date, time, and title) and confirmed upon all attendees' approvals. Coordination means making suggestions for these attributes, either approving or rejecting these suggestions, and giving arguments and counter-arguments before any agreement is reached. Kotlarsky et al. [21] analyze the role of social coordination mechanisms in reducing the knowledge gap between people (e.g., misunderstanding) and creating a social capital (e.g., building trust relationships) for the global team in virtual enterprises. These mechanisms include communication activities, working relationships, and social cognition.

3   Engaging Business- and Social-Data Artifacts in collaboration

In Sect. 1, we mention that enterprise 2.0 is built upon the connection of the business and social worlds together. In this section, we define the constituents of each world in terms of BAs and SAs, revisit the definition of SA presented in [25], and discuss how this connection is shaped through a meet-in-the-middle platform. We illustrate BAs and SAs collaboration with a faculty-hiring scenario.

3.1 Faculty Hiring Running-Example

Knowing that more and more people post their personal profiles (e.g., experience, education, and activities) on social medias[2], many organizations are using these medias as an additional source of information in their hiring endeavors. According to a Forbes post (www.facebook.com/forbes/posts/10153049903042509), 67% of people looking for a job say they use Facebook in their social media search and some 73% of recruiters say they have used some form of social media to hire staff. In addition, "... *The most effective organizations make smart use of employee networks to reduce costs, improve efficiency, and spur innovation*" [9].

Our running example targets universities that could tap into applicants' professional (e.g., LinkedIn and ResearchGate) and social (e.g., Twitter and personal blogs) worlds. Upon receiving applications, a Human Capital (HC) staff screens them for suitability according to the available job postings. One of the common practices, nowadays, is that CVs include URLs to candidates' personal Web pages showing for instance, their academic credentials and professional experiences. Some URLs, also, refer to candidates' social media pages inviting potential universities to check the activities in which they have engaged, such as helping out local communities and supporting special events. The HC staff screens the social pages by pointing his/her browser towards the appropriate URLs. On top of being "painful" and time consuming, this manual screening could overlook some important details if the HC staff has to do it for every candidate.

In this paper, we illustrate how an automatic screening of applications permits to collect details from candidates' professional and social pages. Today's networks like LinkedIn[3] already structure users' profiles in a way that makes their skills, interests, experiences, and educations easy to identify and process. In fact, LinkedIn even provides ways to programmatically access their profiles via online Application Programming Interfaces (API).

3.2 Definition of Social-Data Artifact

Prior to defining SA, we introduce the concept of social action. A social action is an operation that a Web 2.0 application offers to users and other applications (whether Web 2.0 or not) for execution over this Web 2.0 application. Multiple social actions exist like post, invite, tag, and share, and vary from one Web 2.0 application (e.g., post in Facebook) to another (e.g., tweet in Twitter). To keep the paper self-contained, readers can refer to [26] for more details on social actions.

---

[2] "*By the end of 2013, Facebook was being used by 1.23 billion users worldwide, adding 170 million in just one year*" [20].

[3] LinkedIn's list of attributes that support automatic analysis of members' profiles is available at developer.linkedin.com/docs/fields. Examples of attributes include industry, location, specialties, and positions.

SA is a meaningful piece of information that a Web 2.0 application makes available to users and other applications (whether Web 2.0 or not). We define SA from two perspectives:

1. Operation perspective: it associates, in a straightforward manner, a social action with a SA, e.g., post and tag are directly mapped onto postSA and tagSA, respectively.
2. Data perspective: it may associate the result of executing a social action with a SA. Indeed, not all social actions' results should become SAs. While post on Facebook wall creates eventPostSA, sharing this post, that is already associated with eventPostSA, does not create a new SA. Sharing enriches the existing eventPostSA.

Like a BA, a SA includes a set of descriptive data properties (structured and/or unstructured), a set of states that reflect changes in the SA, and a life-cycle built upon these states.

With respect to the faculty hiring example, jobOpeningBA (description of the job opening) and applicationCandidateBA (application of a candidate) are BAs. And, jobOpeningPostSA and jobOpeningTweetSA, that result from executing post on Facebook and tweet on Twitter, respectively, are SAs. Any comment on Facebook will simply enrich jobOpeningPostSA and will not create any new SAs. Table 1 presents the correspondence between BAs and SAs in the context of enterprise 2.0.

Table 1. BA/SA correspondence in the context of enterprise 2.0

Correspondence type	Description
1:0	A BA is not associated with any SA, e.g., candidateSalaryNegotiationBA treated as private with no corresponding SA
1:1	A BA is associated with exactly one SA, e.g., jobOpeningBA associated with jobOpeningPostSA that exists on the university's private social network
1:n	A BA is associated with many SAs, e.g., candidateInformationBA extracted from user profiles that exist on different Web 2.0 applications like userProfileLinkedInSA and userProfileFacebookSA

3.3   Business-Data Artifact *versus* Social-Data artifact

Our vision of enterprise 2.0, thoroughly discussed in [6], advocates for bridging the gap between the business and social worlds using data artifacts. Business-data artifacts reside in the business world and social-data artifacts reside in the social world. To better understand the synergy between these artifacts/worlds, we propose the following five properties (Table 2):

1. Stakeholder refers to entities (e.g., customers and government bodies) that constitute the enterprise's ecosystem. In the business world, the identity of any stakeholder must be known. In the social world, the identity of a stakeholder might not be known, i.e., anonymity is tolerated.
2. Regulation refers to local and international policies that define and constraint the operation of the enterprise. In the business world, compliance with regulations is a must and needs to be enforced at all times. In the social world, compliance with regulations is somehow relaxed. The (physical) locations of both the enterprise and the Web 2.0 application have an impact on this compliance.
3. Environment refers to the ecosystem in which the enterprise operates. This ecosystem could be political like EU (European Union) or economical like BRIC (Brazil, Russia, India, and China). In the business world, the environment is closed by default (i.e., strict control). This is backed by the definitions of stakeholder and regulation properties in the business world. Any business-data artifact should be mapped onto a specific element that is reported in this environment's legislation. In the social world, the environment is open by default (i.e., loose control); e.g., customers enter and leave the enterprise's Facebook account without any prior notice. This is backed by the definition of stakeholder property in the social world.
4. Content refers to data and messages that the enterprise manipulates and exchanges. In the business world, the content is "well structured" although business-data artifacts handle unstructured data. By well structured, we mean that content management operations are properly defined. In the social world, the content is "loosely structured", i.e., different forms of data (e.g., text, photo, and audio files) and messages can be manipulated and exchanged over Web 2.0 applications.
5. Third-party-application refers to the control that the enterprise has over applications of third parties. In the business world, the enterprise has full control over any third-party application that it uses. For instance, it sets its functional and non-functional requirements over this application. In the social world, the enterprise has limited control over third-party applications (i.e., Web 2.0 application like Facebook). For example, the enterprise might not be able to dictate how its account on a Web 2.0 application operates.

Table 2. Properties associated with data artifacts in enterprise 2.0

Property	Business world	Social world
Stakeholder	Identity must be known	Anonymity is tolerated
Regulation	Compliance is a must	Compliance is relaxed
Environment	Closed	Open
Content	Well structured	Loosely structured
Third-party app	Full control	Limited control

3.4   Architecture

Figure 2 represents the architecture of enterprise 2.0. Each world is the host of specific platforms and/or applications for managing business processes and social activities[4]. These activities are initiated when the business processes are executed (in fact, when BAs take on specific states). Interactions between the business and the social worlds involve BAs and SAs. These interactions either are confined into the borders of a specific world or cross the worlds' borders. The interactions between the different artifacts in Fig. 2 are used for representation purposes, only. To bridge the gap between the business and social worlds, a meet-in-the-middle platform is deployed hosting Social Machines (SMs) [7]. SMs act as proxies over Web 2.0 applications so that BAs act upon SAs, when deemed necessary. Figure 2 also shows interactions between known/unknown stakeholders and business/social worlds. The interaction with the business world is for executing business processes by processing known stakeholders' requests and contacting known stakeholders. Meanwhile, the interaction between stakeholders and social world is for using Web 2.0 applications for creating accounts, signing in, posting notes, etc.

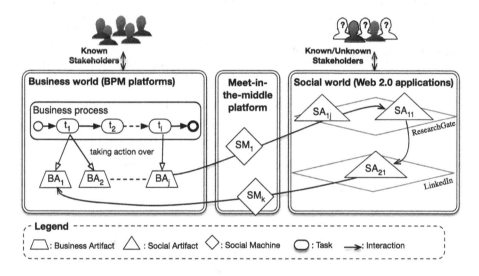

Fig. 2. Architecture of enterprise 2.0

The business world hosts the enterprise's business processes. A business process consists of tasks connected to each other through data dependencies like prerequisite and co-prerequisite. To perform a process, tasks are executed by acting upon the relevant BAs through what is usually known in the database community as Create, Read, Update, and Delete (CRUD) operations. What is

---

[4] Since processes are usually known for being well-formed, we avoid the term of "social processes" and use "social activities" instead. These latter are generally unstructured and unordered.

important in these operations is how a BA takes on specific states according to its respective life-cycle and how a BA, with the help of specific SMs, interacts with the social world to create new SAs, communicate with existing SAs, etc.

The social world hosts the enterprise's SAs associated with Web 2.0 applications and created as per Sect. 3.2. Some applications are internal to the enterprise (i.e., locally managed) while others are external (e.g., LinkedIn) which needs specific agreements (service level agreement) between the enterprise and these applications' providers. An enterprise can consider different vendors of Web 2.0 applications depending on the type of online social presence that it wishes to have on the Internet. We assume that each Web 2.0 application has a set of proprietary APIs that allow third parties to act on this application in terms of creating new SAs, making SAs take on new states, checking on SAs' states, and so on. Like BAs, SAs interact with others in the same or different applications and respond to BAs' queries like informing them about their current states. It is worth noting that CRUD operations are also executed over SAs.

The meet-in-the-middle platform consists of SMs that offer the necessary support for BA/SA interaction. The diversity of Web 2.0 applications (e.g. APIs, protocols, and data format) makes it quite difficult to "consume" their functionalities in a unified way. To deal with this diversity, SMs offer functionalities, exposed as a unified API, that implement these interactions between BAs and SAs. SMs, also, define a generic model for wrapping Web 2.0 applications such as Twitter, Facebook, Google, and Dropbox. Table 3 includes some SMs' functionalities.

Table 3. Social machines' functionalities

Functionality name	From: To	Description
link	B:S	Connects the business world to a platform hosting SAs in the social world
unlink	B:S	Disconnects the business world from a platform in the social world (coupled with *link*)
search	B:S	Looks for SAs according to a provided search query
list	B:S	Identifies SAs according to specific filters, e.g., all users who declined an invitation to a social event
create	B:S	Requests the establishment of a new SA
delete	B:S	Removes a SA (coupled with *create*)
update	B:S	Requests changes in the data/state of a SA
ping	B:S	Asks for details (e.g., current data values and state) of a SA
ack	S:B	Returns the current data values and state of a SA (coupled with *ping*)
subscribe	B:S	Allows the business world to subscribe to an event of interest (e.g., changing in a SA state) and be notified later
unsubscribe	B:S	Allows the business world to unsubscribe of an event of interest (coupled with *subscribe*)
notify	S:B	Notifies the business world when an event of interest occurs in response to *subscribe*

Most messages associated with these functionalities originate from the business world since it hosts the processes that drive the enterprise operation and it has control over the social activities to perform. Since the business and social worlds are associated with different platforms/applications, operation failure can occur. For the sake of simplicity we assume that all operations succeed. Similarly to enterprise portals (e.g., Alfresco and Liferay) that aim at integrating information, people, and processes across organizational boundaries through a unified framework, the meet-in-the-middle platform acts as a liaison between the business and social worlds.

3.5   Illustration

Let us use the faculty-hiring example to illustrate how a SM's functionality like *link* allows the business and social worlds to interact together. First of all, we assume that the university has accounts (usernames/passwords) registered in certain Web 2.0 applications. These accounts enable the HC staff to execute online activities and to generate some data hosted by these Web 2.0 applications. The HC staff signs in into LinkedIn and posts some job openings on the university's LinkedIn page. This is an example of a "manual" creation of posts. However, to allow an automated business process to create posts on the university's LinkedIn page, we need to understand how LinkedIn allows third-party applications to manipulate its functionalities. In practice, this involves authentication to acquire an access token, to request approvals, and to exchange authorization codes prior to calling LinkedIn's management APIs. To allow this automated authentication, a LinkedIn-SM provides a single functionality called *link* (Table 3) that establishes a "pre-authorized" communication channel between the business and social worlds. *link* is responsible for abstracting the whole authentication process.

Figure 3 shows an UML sequence diagram of this process that begins when LinkedIn-SM requests a token from LinkedIn upon a link request from the business world. Next the HC-staff is redirected to a URL along with some necessary input details like login and consent. LinkedIn handles user authentication and consent. The result is an authorization code that LinkedIn-SM exchanges for an access token, which has a limited lifetime, and a refresh token, which allows to obtain new access tokens beyond the lifetime of an access token. LinkedIn-SM then stores the refresh token for future use and uses the token to access LinkedIn APIs. As expected, this process is only performed in the first access. Technical details on the implementation of *link* are given in Sect. 4.

3.6   Examples of Social-Data Artifacts

As stated in Sect. 3.2, a SA is defined from two perspectives: operation (e.g., postSA) and data (e.g., jobOpeningPostSA). These perspectives correspond to the nature of social actions that users execute along with these actions' possible outcomes. In [26], we describe the categorization of social actions into communication, sharing, and enrichment (Table 4). A social action can simultaneously belong to more than one category but this possibility has been discarded for the sake of simplicity.

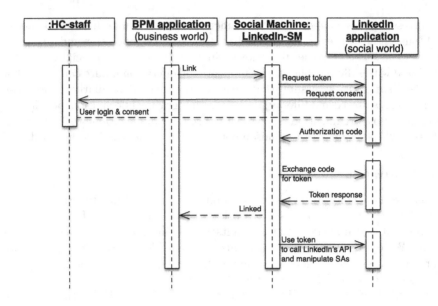

Fig. 3. Authentication using *link* functionality

Table 4. Representative categories of social actions

Category	Description	Examples of social actions
Communication	Includes actions that lead into back-and-forth interactions between users, which should help them engage in joint operations	**Chat** with a user or group of users **Poke** someone **Send** direct messages to a user's inbox **Launch** a social event/campaign
Sharing	Includes actions that allow to create and edit shared content as well as facilitate its consumption; the actions support one-way interaction	**Co-author** a text on a Wiki system **Publish** a post on a Blog Web site **Upload** a video on a public repository **Subscribe** to an RSS Feed
Enrichment	Includes actions that provide additional [meta] data on shared content in terms of providing opinions and/or ranking something	**Comment** a post **Rank** a post, page, video, news **Tag** user's photos, videos, activities **Like/Unlike** a comment or post

In this section, we present some SAs' and BAs' properties and life-cycles. Different BAs and SAs are associated with the faculty-hiring example (Fig. 4). When the university announces a job opening, jobOpeningBA is created so that the opening is tracked from a business perspective e.g., approvals and credentials. jobOpeningBA's life-cycle contains states like created, initiated, and launched. This latter state triggers a post on the university's Facebook account. This post is associated with postSA and jobOpeningPostSA. Values assigned to postSA's properties include (postTrigger: jobOpeningBA) and (postOwner: Mr. Michael from HC) while values assigned to jobOpeningPostSA's properties include post-Trigger: postSA and postOwner: Mr. Michael from HC.

When a candidate accesses the university's Facebook account, he/she can apply for the opening (i.e., create candidateApplicationBA) and/or take further social actions like comment. This leads into creating commentSA and enriching jobOpeningPostSA. Figure 4 illustrates the interactions between some BAs and SAs involved in the faculty-hiring running example.

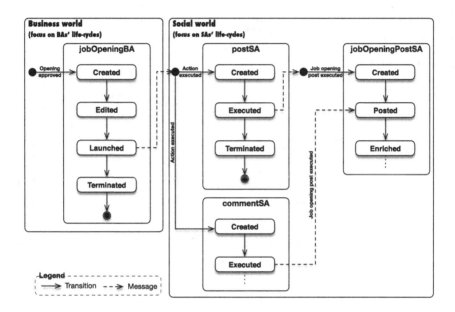

Fig. 4. Artifacts' life-cycles in interaction

4   Faculty-Hiring System Development

We discuss the development of the faculty-hiring system that allows universities to tap into the social world so they identify the right candidates with respect to faculty openings' requirements. We outline the technical architecture of the system as well as the implementation technologies and tools.

4.1   Architecture

Figure 5 illustrates the components of the architecture of the faculty-hiring system. The architecture complies with the design guidelines for enterprise 2.0

presented in [6] by having a meet-in-the-middle platform that bridges the gap between the business and social worlds. The architecture includes a *Data source layer*, a *SA management layer*, a *Business service layer*, and an *Application layer*.

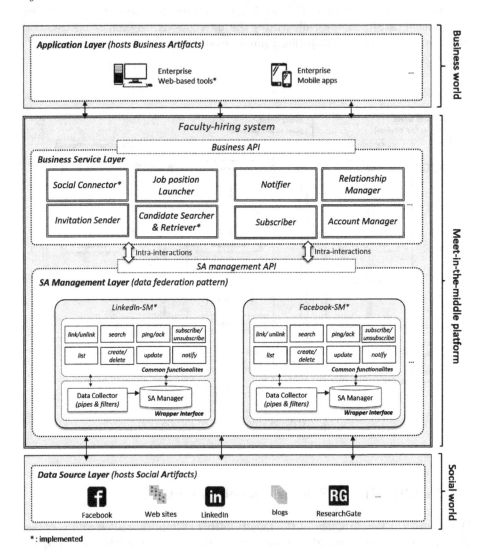

Fig. 5. Faculty-hiring system's architecture

1. *Data source layer* hosts SAs associated with job offers (e.g., jobOpening-PostSA), candidates' profiles (e.g., userProfileLinkedInSA), candidates' comments on openings, etc. SAs are distributed over different Web 2.0 applications such as LinkedIn and Facebook.

2. *SA management layer* hosts specialized social machines like LinkedIn-SM and Facebook-SM for wrapping purposes. Each SM loads, filters, and extracts data of interest from SAs that are afterwards stored in a SA Manager, according to a common SA model. On top of the SA Manager, the functionalities introduced in Table 3 are implemented and provided to the *Business service layer* through SA management APIs.
3. *Business service layer* relies on SA management APIs to define different software components (e.g., *Social connector*, *Job launcher*, and *Notifier*), which implement the business logic of the faculty-hiring system's services. Such services are externalized as REST APIs [16] to be consumed by enterprise business applications.
4. *Application layer* refers to enterprise applications, typically GUIs for the *Business service layer*, that can be developed on top of the faculty-hiring system's services. In Fig. 5, these applications are Web-based and/or mobile-based. In the current version, a Web-based application to use some faculty-hiring system's services is available (youtu.be/Fdvghukdk9w).

4.2 Implementation

Our faculty-hiring system implements the elements marked with asterisk (*) in Fig. 5. The system is implemented in Java 1.8 and deployed on a Tomcat 8.0 server. Facebook and LinkedIn are the Web 2.0 applications included in the *Data source layer* and used for hosting SAs. RESTful style [16] is, also, used to develop both the SA management and Business APIs. Such RESTful APIs are implemented with Jersey (jersey.java.net), which is the JAX-RS (jax-rs-spec.java.ne) reference implementation. The Web-based application consists of a friendly Web interface developed in JSP using Bootstrap (getbootstrap.com) framework. This Web application allows to launch the faculty-hiring system's services in a way to support message exchange between the business and social worlds. Figure 6 shows some screenshots of the developed Web-based application retrieving information on a potential candidate.

On the business world, the system considers jobOpeningBA as the main BA. Meanwhile, on the social world, we highlight some SAs such as jobOpeningPostSA, userProfileLinkedInSA, and userProfileFacebookSA. Figure 7 illustrates the SMs' inner-components in charge of collecting SAs. As shown in this figure, each SM has an internal Data collector that uses pipes and filters as an integration pattern to create the logic for both data conversion and aggregation. Internally, the Data collector is a set of interconnected components that perform tasks such as loading data from the wrapped Web 2.0 application, filtering unnecessary data, and formatting them into the corresponding SA's properties. In practice, as Web 2.0 applications usually provide data in JavaScript Object Notation (JSON, json.org) format, the main conversion taking place within the Data collector is the mapping content from JSON onto SAs and *vice-versa*. The Social Artifact Mapper is the internal component responsible for doing that and making SAs available through the SA manager.

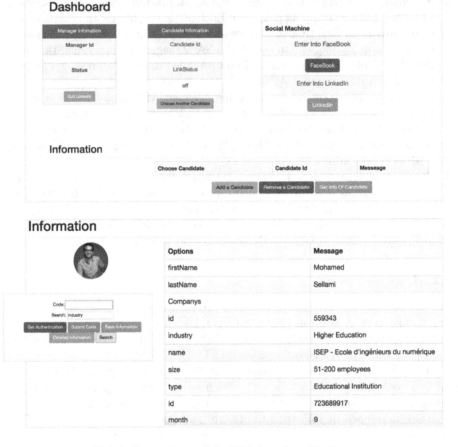

Fig. 6. Screenshots of the Web-based application

As there is a need to integrate heterogeneous SAs from existing Web 2.0 applications, the architecture design of enterprise 2.0 systems encompasses some integration issues. In this sense, it is worth noting that we use the *Data federation pattern* [23] to aggregate and correlate the necessary SAs from multiple Web 2.0 applications. The data federation is realized in the *SA management layer* by a set of parallel *pipes and filters* defined by each individual SM (Fig. 7). *SA Management API* simplifies the manipulation of SAs by abstracting their properties and life-cycles. Furthermore, some generalizations are also implemented in order to extract two or more SAs' shared characteristics and combine them into a unique generalized SA. This generalization process was fundamental to enable a SA common model. For example, candidateProfileSA abstracts both candidateProfileLinkedInSA and candidateProfileFacebookSA.

We comply with REST principles [16] to ensure the proper development of the faculty-hiring system's APIs and allow a seamless access from different

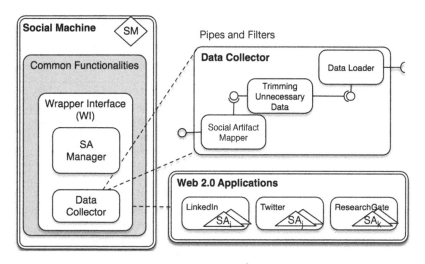

Fig. 7. Internal structure of a SM's data collector (**Z: needs to be fixed**)

platforms including legacy applications. Beyond that, widely known technologies such as JSON for data-interchange format and Open Authorization Protocol (OAuth, oauth.net) for authentication are also used in this development. For instance, *link* functionality is based on OAuth 2.0 that facilitates the establishment of "pre-authorized" and "persistent" communication channels between the business and social worlds, and allows the manipulation of SAs on behalf of the enterprise. JSON is used as a default data format in the interactions between the Web-based application and faculty-hiring system's API. To better understand how the common functionalities already described in Table 3 are implemented as RESTful APIs, Table 5 shows the HTTP method used to execute each of them.

The design and implementation of the faculty-hiring running example points out the degree of integration that can be obtained by using our approach, but also the set of new opportunities that emerge when supporting the connection of the business world and Web 2.0 applications together. One of these opportunities is the adaptation that business processes could be subject ti based on the relevant events that happen in the social world. In the aforementioned SM's functionalities (Table 5), for instance, *notify* could be used to trigger a task of interview scheduling on the business world. This happens when a candidate matches an opening position according to some requirements that could be specified in the *subscribe* functionality.

4.3   Data-Exchange Analysis

In Fig. 5, the meet-in-the-middle platform supports the data exchange between the business and social worlds. From an analysis perspective, we captured the average time and other relevant metrics during this exchange as per Table 6. Our analysis focused on the SM's *retrieve* functionality (Table 5). We assessed the

Table 5. RESTful specification of SM's functionalities

Name	HTTP request	Description
URIs relative to https://{*host*}/{sm-id}		
link	POST /link	Connects the SM to the University's LinkedIn and/or Facebook account
	POST /link/{*candidate-id*}	Connects the SM to the LinkedIn and/or Facebook account of the candidate identified by {*candidate-id*}
unlink	POST /unlink	Disconnects the SM to the University's LinkedIn and/or Facebook account
search	GET /search?*q=query*	Returns candidateProfileSAs related to the search query
list	GET /{*opening-id*}	Gets candidateProfileSAs suitable to the jobOpeningPostSA identified by {*opening-id*}/match
	GET /{*opening-id*}/viewed	Lists candidateProfileSAs that have viewed University's posting on LinkedIn or Facebook about the job opening position identified by {*opening-id*}
	GET /{*opening-id*}/applied	Lists candidateProfileSAs that applied to a job opening position identified by {*opening-id*}
create	POST /{*opening-id*}	Requests the creation of a new jobOpeningPostSA identified by {*opening-id*}
delete	DELETE /{*opening-id*}	Requests the removal of the jobOpeningPostSA identified by {*opening-id*}
update	PUT /{*opening-id*}	Requests changes in the data and/or state of the jobOpeningPostSA identified by {*opening-id*}
retrieve	GET /retrieve/{*id*}	Retrieves basic candidateProfileSA of the candidate identified by {*id*}
	GET /retrieve/{*id*}/detailed	Retrieves detailed candidateProfileSA of the candidate identified by {*id*}
ping	GET /{*opening-id*}	Gets details (e.g., current data values and state) of the jobOpeningPostSA identified by {*opening-id*}
ack	HTTP RESPONSE CODE:200 {*data*}	Returns the current data values and state of a pinged jobOpeningPostSA
subscribe	POST /subscribe/{*topic*}	Subscribes to the *topic* of interest
unsubscribe	POST /unsubscribe/{*topic*}	Unsubscribes to the *topic* of interest
notify	POST {*callbackURL*}	Notifies the business world, via a previously informed callback URL, when an event of interest occurs

response time between our faculty-hiring system and both LinkedIn and Facebook. Retrieving the two types of candidate profiles (i.e., basic and detailed) from Facebook and LinkedIn were simultaneously intercepted. A total of 15,288 requests were tracked over a period of time of 70 min, i.e., 8:40 am–9:50 am (during the regular working hours for an University) with a data calculation every second.

Table 6. Performance analysis of retrieving candidate profiles

	Facebook		LinkedIn	
	Basic profile	Detailed profile	Basic profile	Detailed profile
Number of requests	3809	3824	3828	3827
Average (ms)	273.34	273.63	474.12	480.02
Minimum (ms)	255	255	268	300
Maximum (ms)	1229	1264	4559	4432

From Table 6, we notice that for the same Web 2.0 application, the difference between the average response time for retrieving basic profile and retrieving detailed profile is very small (e.g., Facebook basic profile: 273.34 ms and Facebook detailed profile: 273.63 ms). However, we could realize that LinkedIn presented almost twice the average time assessed with Facebook (e.g., Facebook detailed profile: 273.63 ms; LinkedIn detailed profile: 480.02 ms). Part of this difference is explained by the average amount of data exchanged to get each kind of profile (Table 7).

Table 7. Average amount of data (in Byte) exchanged by each candidate profile

	Facebook	LinkedIn
Basic profile	196 B	284 B
Detailed profile	295 B	500 B

5 Future work

Our future work revolves around 4 initiatives: social query, social-content protection, artifact management, and semantic social machines.

1. In the social-query initiative, we would like to define a language for running queries over social-data artifacts like suggested in [28]. Users could be interested in analyzing the content of some social-data artifacts and identifying patterns between them, for example.

2. In the social-content protection initiative, we would like to address the reluctance of enterprises to embrace Web 2.0 applications. Signs of this reluctance include lack of control over the content posted on these applications and these applications' open nature since members sign up and sign off without prior notice. To motivate the business world's stakeholders embrace Web 2.0 applications, we could think of putting restrictions on social-data artifacts' life-cycles by either enabling or disabling some states.

3. In the artifact-management initiative, we would like to look into ways of applying the five processes of a successful contact center to social-data artifacts [33]. These processes are listening (how to get access to social-data artifacts across various Web 2.0 applications?), funneling (how to find the rights social-data artifacts in response to certain needs?), routing (how to direct relevant social-data artifacts to appropriate stakeholders?), engaging (how to provide stakeholders with the necessary means to track social-data artifacts?), and recording (how to measure the value-added of social-data artifacts to the enterprise operation?).

4. Finally, the semantic social machines initiative would examine the role of ontologies in addressing interaction conflicts that could arise between the business and social worlds. Different terminologies could be used for expressing data and events and hence, consistency would be deemed necessary during interaction.

6   Conclusion

In this paper, we designed and implemented an approach that brings the business and social worlds together. Each world has its priorities and concerns. The approach relies on business-data artifacts, social-data artifacts, and meet-in-the-middle platform. Business-data artifacts capture activities related to business processes, whilst social-data artifacts abstract activities executed over Web 2.0 applications. The meet-in-the-middle platform supports interactions between artifacts through social machines. A running example for faculty hiring was used to illustrate some data artifacts such as jobOpeningBA as a business-data artifact and jobOpeningPostSA as a social-data artifact. Each artifact is a collection of related data (structured and/or unstructured) and has a life-cycle. A comparative analysis between business- and social-data artifacts was also included in this paper using five properties, namely stakeholder, regulation, environment, content, and third-party application.

References

1. Badr, Y., Narendra, N.C., Maamar, Z.: Business artifacts for e-Business interoperability. In: Kajan, E., (ed.) Electronic Business Interoperability: Concepts, Opportunities and Challenges. IGI Global Publishing (2011)

2. Berardi, D., Calvanese, D., De Giacomo, G., Lenzerini, M., Mecella, M.: A foundational vision for e-Services. In: Proceedings of Ubiquitous Mobile Information and Collaboration Systems Workshop (UMICS 2003) held in conjunction with The 15th International Conference On Advanced Information Systems Engineering (CAiSE 2003), Klagenfurt/Velden, Austria (2003)
3. Blinn, N., Lindermann, N., Fäcks, K., Nüttgens, M.: Web 2.0 Artifacts in SME-Networks a qualitative approach towards an integrative conceptualization considering organizational and technical perspectives. In: Proceedings of Software Engineering 2009 (Workshops), Kaiserslautern, Germany (2009)
4. Brambilla, M., Fraternali, P., Vaca, C.: A notation for supporting social business process modeling. In: Proceedings of the Fourth Workshop on Business Process Management and Social Software (BPMS2 2011) held in conjunction with the Seventh International Conference on Business Process Management (BPM 2011), Lucerne, Switzerland (2011)
5. Bruno, G., Dengler, F., Jennings, B., Khalaf, R., Nurcan, S., Prilla, M., Sarini, M., Schmidt, R., Silva, R.: Key challenges for enabling agile BPM with social software. J. Softw. Maintenance Evol. Res. Pract. **23**(10), 297–326 (2011)
6. Burégio, V.A., Maamar, Z., Meira, S.L.: An architecture and guiding framework for the social enterprise. IEEE Internet Comput. **19**(1), 64–68 (2015)
7. Burégio, V.A., Meira, S.L., Rosa, N.S., Garcia, V.C.: Moving towards "Relationship-aware" applications, services: a social machine-oriented approach. In: Proceedings of the EDOC 2013 Workshops held in conjunction with the 17th IEEE International Enterprise Distributed Object Computing Conference (EDOCW 2013), Vancouver, BC, Canada (2013)
8. Busch, P., Fettke, P.: Business process management under the microscope: the potential of social network analysis. In: Proceedings of the 44th Hawaii International Conference on System Sciences (HICSS 2011), Kauai, HI, USA (2011)
9. Cross, R., Gray, P., Cunningham, S., Showers, M., Thomas, R.J.: The collaborative organization: how to make employee networks really work. MITSloan Manag. Rev. **51**(1), 83–90 (2010). Fall
10. Damaggio, E., Hull, R., Vaculín, R.: On the equivalence of incremental and fixpoint semantics for business artifacts with guard-stage-milestone lifecycles. Inform. Syst. **38**(4), 561–584 (2013)
11. Dengler, D., Koschmider, A., Oberweis, A., Zhang, H.: Social software for coordination of collaborative process activities. In: zur Muehlen, M., Su, J. (eds.) BPM 2010. LNBIP, vol 66, pp. 396–407. Springer, Heidelberg (2010). doi:https://doi.org/10.1007/978-3-642-20511-8_37
12. Draheim, D.: Social BPM, chapter Smart Business Process Management. Future Strategies (2012)
13. Dumas, M., Hull, R., Patrizi, F., Editorial, G.: Special issue on data and artifact-centric business processes. Comput. **98**(4), 343–344 (2016)
14. Erol, S., Granitzer, M., Happ, S., Jantunen, S., Jennings, B., Koschmider, A., Nurcan, S., Rossi, D., Schmidt, R., Johannesson, P.: Combining BPM and social software contradiction or chance. J. Softw. Maintenance Evol. Res. Pract. **22**(6–7), 449–476 (2010)
15. Faci, N., Maamar, Z., Kajan, E., Benslimane, D.: Research roadmap for the enterprise 2.0 issues & solutions. Scientific Publications of the State University of Novi Pazar Journal, Ser. A: Appl. Math. Inform. & Mechanics **6**(2), 81–89 (2014)
16. Fielding, R.T.: Architectural styles and the design of network-based software architectures. Ph.D. thesis, University of California, Irvine (2000)

17. Gonzalez, P., Griesmayer, A., Lomuscio, A.: Verifying GSM-based business arti-facts. In: Proceedings of the 19th IEEE International Conference on Web Services (ICWS 2012), Honolulu, HI, USA (2012)
18. Grim-Yefsah, M., Rosenthal-Sabroux, C., Thion, V.: Using information of an infor-mal network to evaluate business process robustness. In: Proceedings of the Inter-national Conference on Knowledge Management and Information Sharing (KMIS 2011), Paris, France (2011)
19. Houston, J., Hoehler, J.: The Social Enterprise: What it means to be a Social Enterprise, Deloitte Digital (2013). http://deloitteblog.co.za/wp-content/uploads/downloads/2013/04/Enterprise-2.0.pdf
20. Kiss, J.: Facebook's 10th Birthday: From College Dorm to 1.23 Bil-lion Users (2014). http://www.theguardian.com/technology/2014/feb/04/facebook-10-years-mark-zuckerberg
21. Kotlarsky, J., van Fenema, P.C., Willcocks, L.P.: Developing a knowledge-based perspective on coordination: the case of global software projects. Inform. Manag. **45**(2), 96–108 (2008)
22. Kumaran, S., Liu, R., Wu, F.Y.: On the duality of information-centric and activity-centric models of business processes. In: Proceedings of the 20th International Con-ference on Advanced Information Systems Engineering (CAiSE 2008), Montpellier, France (2008)
23. Liu, Y., Liang, X., Xu, L., Staples, M., Zhu, L.: Using architecture integration patterns to compose enterprise mashups. In: Proceedings of the 8th Working IEEE/IFIP Conference on Software Architecture (WICSA 2009) and the 3rd Euro-pean Conference on Software Architecture (ECSA 2009), Cambridge, UK (2009)
24. Maamar, Z., Badr, Y., Narendra, N.C.: Business artifacts discovery and modeling. In: Proceedings of the 8th International Conference on Service Oriented Computing (ICSOC 2010), San Francisco, California, USA (2010)
25. Maamar, Z., Burégio, V., Sellami, M.: Collaborative enterprise applications based on business and social artifacts. In: Proceedings of the 24th IEEE International Conference on Enabling Technologies: Infrastructure for Collaborative Enterprises (WETICE 2015), Larnaca, Cyprus (2015)
26. Maamar, Z., Burégio, V.A., Faci, N., Benslimanem, D., Sheng, Q.Z.: "Controlling" Web 2.0 applications in the workplace. In: Proceedings of the 2015 IEEE 19th International Conference on Enterprise Distributed Object Computing (EDOC 2015), Adelaide, Australia, September 2015
27. Masli, M., Geyer, W., Dugan, C., Brownholtz, B.: The design and usage of tentative events for time-based social coordination in the enterprise. In: International World Wide Web Conference, Hyderabad, India (2011)
28. Montemayor, J., Diehl, C.P., Query, S.: Looking for social signals from online artifacts. Johns Hopkins APL Tech. Digest **30**(1), 41–46 (2011)
29. Nandi, P., Kumaran, S.: Adaptive business objects - a new component model for business integration. In: Proceedings of the International Conference on Enterprise Information Systems (ICEIS 2005), Miami, Florida, USA (2005)
30. Narendra, N.C., Badr, Y., Thiran, P., Maamar, Z.: Towards a unified approach for business process modeling using context-based artifacts and web services. In: Proceedings of the 2009 IEEE International Conference on Services Computing (SCC 2009), Bangalore, India (2009)
31. Nigam, A., Caswell, N.S., Artifacts, B.: An approach to operational specification. IBM Syst. J. **42**(3), 415–432 (2003)

32. OpenKnowledge. Social Business Process Reengineering. Technical report (2012), http://socialbusinessmanifesto.com/social-business-process-reengineering (checked out in May 2016)
33. Oracle. Is Social Media Transforming Your Business? An Oracle White Paper, March 2012. http://www.oracle.com/us/products/applications/social-trans-bus-wp-1560502.pdf. visited June 2013
34. Popova, V., Fahland, D., Dumas, M.: Artifact lifecycle discovery. Int. J. Coop. Inform. Syst. **24**(1), 1–44 (2015)
35. Rito Silva, A., Meziani, R., Magalhães, R., Martinho, D., Aguiar, A., Flores, N.: AGILIPO: embedding social software features into business process tools. In: Proceedings of the Third International Workshop on the Business Process Model and Notation (BPMN 2010), Ulm, Germany (2010)
36. Ryndina, K., Malte Küster, J., Gall, H.: Consistency of business process models and object life cycles. In: Proceedings of Models in Software Engineering Workshops (MoDELS 2006), Genova, Italy (2006)
37. Strosnider, J.K., Nandi, P., Kumaran, S., Ghosh, S., Arsanjani, A.: Model-driven synthesis of SOA solutions. IBM Syst. J. **47**(3), 415–432 (2008)
38. Warr, W.A., Software, S.: Fun and games, or business tools? J. Inform. Sci. **34**(4), 591–604 (2008)
39. www.lightwaveonline.com. Web 2.0 Companies Likely Shape Optical Network Market, March 2015. www.lightwaveonline.com/articles/2015/03/web-20-companies-likely-shape-optical-network-market-says-lightcounting.html (checked out in May 2016)
40. www.strategyr.com. Global Enterprise Web 2.0 to Reach US$5.7 Billion by 2010, June 2015. www.strategyr.com/PressMCP-6116.asp (checked out in May 2016)

Privacy-Preserving Querying on Privately Encrypted Data in the Cloud

Feras Aljumah[1,2(✉)], Makan Pourzandi[3], and Mourad Debbabi[1]

[1] Concordia University, Montreal, QC, Canada
[2] Al-Imam Mohammad Ibn Saud Islamic University, Riyadh, Saudi Arabia
faaljumah@imamu.edu.sa
[3] Ericsson Research Security, Montreal, QC, Canada

Abstract. Cloud services provide clients with highly scalable network, storage, and computational resources. However, these service come with the challenge of guaranteeing the confidentiality of the data stored on the cloud. Rather than attempting to prevent adversaries from compromising the cloud server, we aim in this paper to propose a protocol for secure querying in the cloud, while preserving the privacy of the participants and assuming the existence of a passive adversary able to access all data stored in the cloud. In this paper, we address this problem by proposing a network protocol that would allow a third party, such as a health organization, to query privately encrypted data without relying on a trusted entity. The protocol we propose preserves the privacy of the data owners and the querying entity. The protocol relies on homomorphic cryptography, threshold cryptography, differential privacy, and randomization to allow for secure, distributed, and privacy-preserving queries. We evaluate the performance of our protocol and report on the results of the implementation.

1 Introduction

In recent years, storing data on the cloud has become popular due to it simplicity, integration, affordability, and the wide range of features offered by service providers. However, there are many applications, where the privacy of the data remains a concern due to the sensitive nature of the data. Such systems include cloud-based personal health record (PHR) or financial management systems. In the market today, there are many solutions which have been presented to provide users with cryptographically secure storage. These systems guarantee that access is only possible through the user's encryption keys, which are not accessible or stored by the cloud provider. These solutions provide data confidentiality, but also prevent the system from generating useful reports or statistics based on users' data.

PHR systems are patient centric systems that allow patients to store and manage health data stored on the cloud. Many of the current PHR system providers have the ability to access all patient records [1]. Even though PHRs might be encrypted in the cloud, the keys are being managed by the same provider. As such,

© Springer-Verlag GmbH Germany 2017
A. Hameurlain et al. (Eds.): TLDKS XXXV, LNCS 10680, pp. 50–73, 2017.
https://doi.org/10.1007/978-3-662-56121-8_3

it is possible for intruders with access to the cloud provider's infrastructure to gain access to all the records. This led researchers to propose solutions that enforce access control to PHRs using Attribute Based Encryption (ABE) [2]. ABE is utilized to encrypt and store PHR data on semi-trusted servers using access control policies chosen by patients [3, 4]. In such a system, it is assumed that no user should have access to all PHRs.

Although the use of an ABE system preserves the privacy of the users, it prevents third party entities such as health organizations from querying privately encrypted data (e.g. PHRs) on the system. To produce statistical information on privately encrypted data on the cloud, users would have to give the query requester access to all private data using ABE. According to a report from the consulting firm PwC [5], health organizations are falling short in protecting the privacy and security of patient information. Additionally, according to the same report, more than half of health organizations have had at least one issue with information security and privacy since 2009. The most frequently observed issue is the improper use of protected health information by an employee in the organization.

Solutions such as CryptDB [6] and Monomi [7] have been proposed to execute SQL queries over encrypted data. We show in the related works section the differences between these solutions and our proposed solution and show why they can not be used to solve the problem we described above.

We therefore propose a solution to the problem of querying privately encrypted data under the assumption of not having a trusted entity in the system. This implies that the user trusts neither the system nor the querying entity (QE) to have direct access to the data. This also means that the QE trusts neither the system nor the users to see the query details in the query. The protocol also ensures that the Personally Identifiable Information (PII) of the data owners is protected. We achieve this by means of a protocol which uses private comparison protocols along with semantically secure cryptography to compare the encrypted values. To protect (PII) we use differential privacy to add noise to the final result. This allows us to support equality, range, and aggregate queries such as *count*, *sum*, and *average*. We also use threshold Goldwasser-Micali cryptography to prevent the QE or the cloud server (CS) from viewing the actual user data. Finally, we use the randomization and partial decryption of the results array to prevent the CS from correlating a result to a specific user.

This paper is an extension of work originally reported in [8]. In this paper, we increase the security of our previously proposed protocol by addressing two additional threat models, which we present in Sect. 2.4 and address in Sects. 5 and 6.

Variations of our approach may solve a variety of problems other than the PHR example mentioned in the introduction, where a party needs to query multiple data sources that are privately encrypted while neither party trusts the other with any data aside from the final results of the query. Examples of some possible use cases include:

- Credit card companies querying private e-commerce data to find cases of credit card fraud or identity theft.
- Insurance providers querying pharmacies for illegal abuse by patients, doctors, or pharmacists related to drug prescriptions.

The remainder of the paper is organized as follows. In Sect. 2, we present the threat model and the security guarantees. In Sect. 3, we give a technical background on the cryptographic primitives used in our protocol. Section 4 describes our protocol in detail. Section 5 describes an extension to the protocol to hide the attributes of the query. Section 6 present our proposed solution to protect the (PII) of the data owners. Section 7 provides a security analysis of our protocol. Section 8 presents the results of our implementation. The related works are presented in Sect. 9.

2 Problem Formulation

2.1 Security Overview

Figure 1 shows the architecture of our proposed solution. Our solution works by outsourcing the homomorphic computations to a semi-trusted cloud server (CS). The QE encrypts the query constants to preserver its privacy. Threshold encryption is used to protect the confidentiality of the QE's query constants. Threshold encryption along with randomization preserves the privacy of the data owners'. The CS runs all the homomorphic computations on the encrypted data, and it only sends the final result back to the QE. This ensures that our system guarantees the confidentiality of the QE's query constants and the Data Owners' data. The solution does not provide confidentiality if the QE and the CS collude and share their encryption keys.

Although our solution guarantees data confidentiality, it does not ensure the integrity, freshness, or completeness of the results returned to the QE. An adversary that compromises the cloud server or a data owner can modify the data and the final result. We now describe the entities in our system, the threat model, and the security guarantees provided under the threat model.

2.2 Entities

There are four main entities in our protocol (as shown in Fig. 1):

- *Data Owners (DO)*, who own the data and want to store it in a cloud server while keeping it confidential from the cloud server and Querying Server.
- *Cloud Storage Servers*, which are used by the DOs to store their encrypted data which are confidential from the cloud service providers. In our protocol, DOs' data may be stored with different cloud providers, under different standards, or encrypted using different cryptosystems. We are using the cloud storage servers as an example, although the privately encrypted data may be stored locally by the DO on an offline storage device.

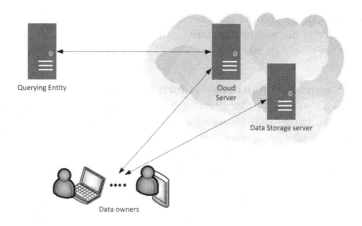

Fig. 1. System architecture

- *Cloud Server (CS)*, which is an intermediary entity in the cloud between the DOs and the QE, trusted to help manage the execution. The DOs and the QE trust the CS to execute the queries, but neither entities trust the CS to see unencrypted data.
- *Querying Entity (QE)*, which runs queries over the encrypted data of the DOs to produce statistical data.

2.3 Problem Statement

Given a set of n data owners $DO\{DO_1, DO_2, \ldots, DO_n\}$, each storing their privately encrypted data d_i on an offline storage device or on a semi-trusted cloud with different encryption keys or cryptosystems. An external querying entity QE is securely allowed to run a query Q with a set of m attributes $Q_{a_i} \in \{Q_{a_1}, Q_{a_2}, \ldots, Q_{a_m}\}$, $i = 1, \ldots, m$ and a set of j constants $Q_{c_i} \in \{Q_{c_1}, Q_{c_2}, \ldots, Q_{c_j}\}$, $i = 1, \ldots, j$, on the encrypted data d_i while ensuring the confidentiality of the QE's query constants Q_c. The QE cannot ascertain any information about the DOs' data other than the final result Q_r and the number of DOs that satisfy the QE's query Q.

2.4 Threat Models

Threat 1: Compromising the Survey Server or the Querying Entity. In this threat, our solution guards against a curious cloud server, a curious querying entity, or other external attackers with full access to either server. Our goal is confidentiality (data secrecy), not integrity or availability. Since we are assuming the semi-honest model, the attacker is assumed to be passive, which means that the attacker wants to learn about the confidential data in the querying entity's query, the data owners' responses, or the final result. The semi-honest model implies that the attacker is not able to modify the data in the query, the data owner responses, or the computation results. This threat is becoming increasingly important with the increasing popularity of migrating data centers to the cloud. We address this threat in Sect. 7.

Threat 2: Personally Identifiable Information Queries. In this threat, the querying entity's goal is to take advantage of the secure querying mechanism to create targeted queries that expose Personally Identifiable Information (PII). For example, a malicious QE may create a query with the goal of finding out if a specific person has cancer. By requesting a **count** query for the number of data owners with "diabetes", living in "location" x, are y years old, and are z cm tall. In most cases, a query this specific would usually be unique to a single DO. Our solution guards against such attacks and ensures the data owners that their PII would not be exposed to a querying entity. Our system achieves this while providing the querying entity with an approximate result using differential privacy. We address this threat in Sect. 6.

Threat 3: Exposing the Query Attributes to the Data Owners. In this threat, our solution guards against a DO attempting to learn about the attributes in the QE query. To address this threat, our solution hides the query attributes along with the query predicates from the data owners. If the QE were to request a query after a major health outbreak, then exposing the attributes in the query might cause DOs to panic. For Example, if a new fatal virus was to spread and a health organization chose to query asking about DOs' with "Diabetes" and "Pregnancy" attributes, then that might cause all DOs with Diabetes and Fever to panic and unnecessarily flood hospitals for unneeded test. The initial solution we present in Sect. 4 hides the query predicates from the data owners to prevent the data owners from knowing the specifics of what the querying entity is interested in. We chose not to hide the query attributes in the initial solution because one of our goals is to return control of data to the DOs themselves. DOs should have the right to prevent the querying of any attribute or any combination of attributes of their data. In Sect. 5, we present a second solution as an extension to the initial protocol to address this threat by hiding the attributes from the DOs.

3 Cryptographic Primitives

In this section, we present an overview of the building blocks that will be utilized in this paper.

Homomorphic Encryption. This is a form of encryption where a specific algebraic operation performed on plaintext is equivalent to another (possibly different) algebraic operation performed on ciphertext. The Paillier's protocol [9] is an additive homomorphic public key encryption. Using Paillier's protocol, given two ciphertexts $E(x)$ and $E(y)$, an encryption of their sum $E(x+y)$ can be efficiently computed by multiplying the ciphertexts modulo a public key N, i.e.,

$E(x + y) = E(x).E(y) \bmod N$. The Goldwasser-Micali (GM) cryptosystem is a semantically-secure protocol based on the quadratic residuosity problem [10]. It has XOR homomorphic properties, in the sense that $E(b) \cdot E(b') = E(b \oplus b') \bmod N$, where b and b' are bits and N is the public key. Variations of the homomorphic Paillier and GM cryptosystems are the distributed threshold decryption protocols in which the decryption is performed by a group of participants rather than one party [11,12]. In this case, each participant would obtain a share of the secret key by executing the distributed key generation algorithm detailed in [13].

Private Comparisons. Yao's classical millionaires' problem [14] involves two millionaires who wish to know who is richer. However, they do not want to inadvertently find out any additional information about each other's wealth. More formally, given two input values x and y, which are held as private inputs by each party respectively, the problem is to securely evaluate the Greater Than (GT) condition $x > y$ without exposing the inputs. In this paper, we employ Fischlin's protocol [15] for private comparison because it allows us to compare two ciphertexts encrypted with the GM cryptosystem using the same public key. Fischlin's protocol takes as input two ciphertexts encrypted using the GM cryptosystem and produces ciphertext sequences, namely Δ and c, that are encrypted by the same public key. The decryption of these sequences would reveal the result of comparing the private inputs without revealing anything beyond the result of the comparison. Fischlin's protocol utilizes the XOR-homomorphic GM cryptosystem to privately compute:

$$x > y \iff \bigvee_{i=1}^{n} \left(x_i \wedge \neg y_i \wedge \bigwedge_{j=i+1}^{n} (x_j = y_j) \right)$$
$$\iff \bigoplus_{i=1}^{n} \left(x_i \wedge \neg y_i \wedge \bigwedge_{j=i+1}^{n} \neg (x_i \oplus y_i) \right)$$

where $|x| = |y| = n$.

4 Proposed Protocol

The purpose of this approach is to propose a protocol that enables querying privately encrypted data on a semi-trusted cloud while preserving the privacy of both the DOs and the QE without requiring a trusted entity.

The protocol uses the Goldwasser-Micali (GM) cryptosystem along with Fischlin's protocol to enable DOs to securely compare values in their data to the encrypted constants in the QE's query.

The protocol additionally uses threshold encryption to allow the QE and the CS to cooperate in order to execute the query without compromising the privacy of either the QE or the DO. Threshold encryption allows us to split an encryption or decryption key into n shares, where only k of the n shares are needed to encrypt or decrypt a message [16]. Threshold encryption allows the CS to calculate the response to the query over encrypted data without trusting it to see the query constants, the DO data, or the result of the query.

4.1 Notations

Throughout this paper, we will be using the following notations:

- QE: Querying entity.
- CS: Cloud server.
- Q: Query sent by QE with the query attributes and the encrypted constants.
- Q_c: Query constants.
- Q_a: Query attributes.
- PK_{gm_i}: The ith private GM-threshold key share.
- pk_{gm}: Public GM-threshold key.
- PK_{gm_i}: The ith private Paillier-threshold key share.
- pk_{pa}: Public Paillier-threshold key.

4.2 Functions

Throughout this paper, we will be using the following functions:

- $PaillierEnc_{pk}(m)$: Encrypts a message m using the Paillier cryptosystem and the session public key [9].
- $GmEnc_{pk}(m)$: Encrypts a message m using the GM cryptosystem and the session public key [10].
- $PaillierSum\,(a, b)$: Runs the Paillier secure sum algorithm adding the two encrypted values a and b, where a and b are both encrypted with the same key using the Paillier cryptosystem [9]. The function outputs an encrypted value c, which is also encrypted with the same key used for a and b.
- $RunFischlin\,(a, b)$: Runs Fischlin's algorithm to compare a and b, which are both encrypted with the same key and using the GM cryptosystem [15]. The functions outputs two values $(\Delta,\,c)$, which are also encrypted with the same key used for a and b. Decrypting $(\Delta,\,c)$ allows us to analyze their values to determine whether a is greater than or equal to b.
- $FischlinResAnalysis(\Delta, c)$: Analyzes the decrypted values of (Δ, c) to determine the comparison result of the two input values to the $RunFischlin\,(a, b)$ function that was used to generate $(\Delta,\,c)$ [15].
- $PaillierDecrypt(m, PK_a)$: Runs the Paillier-threshold decryption algorithm to partially decrypt the message m using a key share a of the private key PK [9].
- $GMDecrypt(m, PK_a)$: Runs the GM-threshold decryption algorithm to partially decrypt the message m using a key share a of the private key PK [10].

4.3 Protocol Phases

The protocol is split into four main phases:

- **Setup phase:** In this phase, the CS and QE agree on the session encryption keys required for the QE's query.

- **Query distribution phase:** In this phase, the QE encrypts the queries and sends them to the CS. The CS then processes the queries and forwards them to the DOs.
- **Data Owner Query execution phase:** In this phase, the DOs run the comparison and sum algorithms on the QE's data and their own encrypted data, and send the results to the CS.
- **Cloud server query execution phase:** In this phase, the CS analyzes and calculates the final result of the query with the help of the QE.

The protocol differs in its operations according to the query type. In other words, the overall structure stays the same; however, there are three deviations according to three query types (*Sum/Avg*, *Range*, and *Hybrid* queries).

4.4   Setup Phase

This section discusses how the QE and the CS start a new query session in our protocol using the GM-threshold cryptosystem along with Fischlin's protocol.

In the first step of the protocol, the secure key agreement protocol is run jointly by the QE and the CS to generate the session public key pk for $k - out - of - n$ threshold GM cryptosystem or the Pailler cryptosystem such that at least k of n participants are required to fully decrypt a ciphertext [11,15]. In this solution, we set the threshold k to 2 and the number of participants n to 2. The generated threshold keys will be kept privately by the QE and the CS.

The QE sends a query session request to the CS along with the type of query (range, sum, or hybrid). The identities of the QE and the CS must be verified at this stage (e.g. by using digital signatures). The type of query determines which keys are generated using the secure key agreement protocol as follows:

- For sum or average queries, the QE and CS generate a Paillier-threshold key pair, splitting the private key PK_{pa} into two shares PK_{pa1} and PK_{pa2}. The key PK_{pa1} will be used by the QE and the key PK_{pa2} will be used by the CS.

Phase 1. Sum/Avg query example

To find the average age of DOs in the system, the following query Q will be sent by the QE to the CS:
$AVG\,(age)\,from\,DOs$

- For range queries, which include queries such as equality, range, and Max/Min queries, the QE and CS generate a GM-threshold key pair, splitting the private key PK_{gm} into two shares PK_{gm1} and PK_{gm2}. The key PK_{gm1} will be used by the QE and the key PK_{gm2} will be used by the CS.

Phase 1. Range query example

To find the number of DOs under the age of 20 that have diabetes, the following query Q will be sent by the QE to the CS:

$COUNT\,(*)\,from\,DOs\,WHERE\,age < (Enc_{pk_{gm}}(20) \wedge Diabetes = Enc_{pk_{gm}}(1))$

- For hybrid queries with range and sum operations, the QE and CS generate a GM-threshold and a Paillier-threshold key pair, splitting each private key into two shares.

Phase 1. Hybrid query example

To find the average age of female DOs with diabetes in the system, the following query Q will be sent by the QE to the CS:

$AVG(age)\,from\,DOs\,WHERE\,gender = Enc_pk_{gm}(1)\,AND\,Diabetes = Enc_pk_{gm}(1)$

The session public keys pk_{gm} for the GM-Threshold cryptosystem and pk_{pa} for Paillier-Threshold cryptosystem will be separately signed by both parties using their trusted certificates before being sent to the DOs with the queries in the upcoming steps of the protocol.

4.5 Query Distribution Phase

The QE encrypts the constants Q_c in the query Q using the session public key pk_{gm} or pk_{pa} depending on the type of query. The QE then forwards the query to the CS.

The CS then distributes the attributes Q_a in the query Q, the encrypted constants Q_c, the public session key pk_{gm} or pk_{pa}, and the cryptosystem to be used (*Paillier-threshold* or *GM-threshold*) to all the DOs.

- In the case of sum or average queries, the DOs would be requested to use the *Paillier cryptosystem*.

Phase 2. Sum/Avg queries

1: $QE \rightarrow CS : Q$
2: $CS \rightarrow DOs :\ Q_a[\,];\ Paillier;\ pk_{pa}$

- In the case of range queries, the DOs would be requested to use the *GM cryptosystem*.

Phase 2. Range queries

1: $QE \rightarrow CS : Q$
2: $CS \rightarrow DOs :\ Q_a[\,];\ GMEnc_{pk_{gm}}(Q_c[\,]);\ GM;\ pk_{gm}$

– In the case of a hybrid query, both of the above actions are required. The CS sends two messages to the DOs: one for the comparison part of the query and another for the Avg/Sum part of the query.

Phase 2. Hybrid queries

1: $QE \rightarrow CS : \{Q\}$
2: $CS \rightarrow DOs : \{ Q_a[\,]; \ Paillier; \ pk_{pa}\}$
3: $CS \rightarrow DOs : \{ Q_a[\,]; \ GMEnc_{pk_{gm}}(Q_c[\,]); \ GM, \ pk_{gm}\}$

The queries are not sent to the DOs, but the attributes and the encrypted constants are sent in arrays of the same size. In the case of comparison queries, the constants in the array correspond to the attributes in the same positions in the attributes array.

4.6   Data Owner Query Execution Phase

The DOs retrieve the values related to the attributes in the QE's query from the cloud.

– If the DOs are requested to use the Paillier cryptosystem, they encrypt the values retrieved from the cloud and encrypt them using the Paillier cryptosystem. The final encrypted result is then forwarded to the CS.

Phase 3. Sum/Avg queries

1: $for \ each \ Q_a \ do \ \{$
   $results_{pa}[i] \ = \ Enc_{pk_{pa}}(DOdata.Q_{a_i})) \ \}$
2: $DOs \rightarrow CS : results_{pa}[\,]$

– If the DOs are requested to use the GM cryptosystem, they encrypt the values retrieved from the cloud and run Fischlin's protocol on their values and the encrypted values sent from the QE. The DOs then forward the results (Δ, c) to the CS.

Phase 3. Range queries

1: $for \ each \ Q_a \ do \ \{$
   $results_{gm}[i] \ = \ RunFischlin(GMEnc_{pk_{gm}}(Q_{c_i}), \ GMEnc_{pk_{gm}}(DOdata.Q_{a_i})) \ \}$
2: $DOs \rightarrow CS : results_{gm}[\,]$

– In the case of a hybrid query, both of the above actions are required, and the DOs send the CS the final two arrays $results_{gm}$ and $results_{pa}$.

4.7   Cloud Server Query Execution Phase

The CS stores the results received from the DOs into an array and waits until it receives the required number of results from the DOs. Depending on the query type, the CS then takes the following steps:

– In the case of a sum/average query, the CS runs the addition algorithm on all the Paillier encrypted values received from the DOs. The CS then partially decrypts the result using its share of the private key. The partially decrypted result is then sent to the QE along with the number of DOs. The QE then decrypts the results using its private key share. If the query was an average query, the QE also divides the result by the number of DOs.

Phase 4. Sum/Avg queries

1: $CS : EncResult = PaillierSum(results_{pa}[\,])$
2: $CS : PartiallyDecResult = PaillierDecrypt(EncResult; PK_{pa_2})$
3: $CS \rightarrow QE : PartiallyDecResult; NumberOfDOs$
4: $QE : FinalResult = PaillierDecrypt(PartiallyDecResult; PK_{pa_1})$
   If average is needed:
5: $QE : Avg = FinalResult/NumberOfDOs$

– In the case of range queries, all results are sent to the QE in an array. The QE then randomizes the order of the elements in the array. The purpose of the randomization by the QE is to prevent the CS from correlating the result of any comparison to a specific DO. The QE then decrypts all the elements in the array using its share of the private key PK_{gm_1}. The new array is then sent to the CS.

The CS then uses its share of the private key to complete the decryption of the $(\varDelta, c)$ elements in the array. The CS can then use $(\varDelta, c)$ to determine the results of the comparisons between the encrypted values in the DO's data and the encrypted constants Q_c in the QE's query. The QE then counts the number of results that satisfy the conditions in the QE's query. The final result is then sent to the QE.

Phase 4. Range queries

1: $CS \rightarrow QE ; results_{gm}[\,]$
2: $QE : PartiallyDecResult[\,] = GMDecrypt(Rand(results_{gm}[\,]); PK_{gm_1})$
3: $QE \rightarrow CS ; PartiallyDecResult[\,]$
4: $CS : DecryptedResult[\,] = GMDecrypt(PartiallyDecResult[\,]; PK_{gm_2})$
5: $CS : if(FischlinResAnalysis(DecryptedResult[i]))\ counter + +;$
6: $CS \rightarrow QE ; counter$

– In the case of a hybrid query, the CS sends the GM-threshold and Paillier-threshold encrypted result arrays to the QE. The QE then randomizes the order of the elements in both arrays. The same randomization order is applied to both arrays to maintain the relationships between them. The QE then decrypts all the elements in the GM-threshold array using its share of the private key. The QE then chooses a new random value, encrypts it using the public Paillier key, and runs the homomorphic addition algorithm on all the values in the Paillier encrypted results array. The new arrays are then sent to the CS. The CS then uses its share of the private key to complete the decryption of the (Δ, c) elements in the GM-threshold encrypted array.

The CS can then use (Δ, c) to determine the results of the comparisons between the encrypted values in the DO's data and the encrypted values sent by the QE. The CS then runs the addition homomorphic protocol on the values in the Paillier array, which are in the positions of the value that satisfy the condition in the GM-threshold encrypted array. The CS then partially decrypts the encrypted sum result using its share of the Paillier private key. The CS then sends the QE the partially decrypted sum result along with the number of DOs, which satisfy the query condition if the QE's query required the average value. The QE then decrypts the result using its private key share and deducts the random value from the result, which is the random value multiplied by the number of DOs. If the query was an average query, the QE also divides the result by the number of DOs. The flowchart in Fig. 2 shows the complete process of executing a hybrid query.

Phase 4. Hybrid queries

1: $CS \rightarrow QE : results_{gm}[\,]; results_{pa}[\,]$
2: $QE : PaillierSum(rand; results_{pa}[\,])$
3: $QE : Randomize(results_{gm}[\,]; results_{pa}[\,])$
4: $QE : PartiallyDecGMResult[\,] = GMDecrypt(results_{gm}[\,]; PK_{gm_1})$
5: $QE \rightarrow CS : PartiallyDecGMResult[\,]; results_{pa}$
6: $CS : DecryptedResult[\,] = GMDecrypt(PartiallyDecGMResult[\,]; PK_{gm_2})$
7: $CS : if(FischlinResAnalysis(DecryptedResult[i]))$
8:    $\{\ sumResult + = results_{pa}[i];$
9:      $counter + +; \}$
10: $CS : PartiallyDecResult = PaillierDecrypt(sumResult; PK_{pa_2})$
11: $CS \rightarrow QE : PartiallyDecResult; counter$
12: $QE : FinalResult = PaillierDecrypt(PartiallyDecResult; PK_{pa_1}) - (rand *$
    $counter)$
    If average is needed:
13: $QE : Avg = FinalResult/counter$

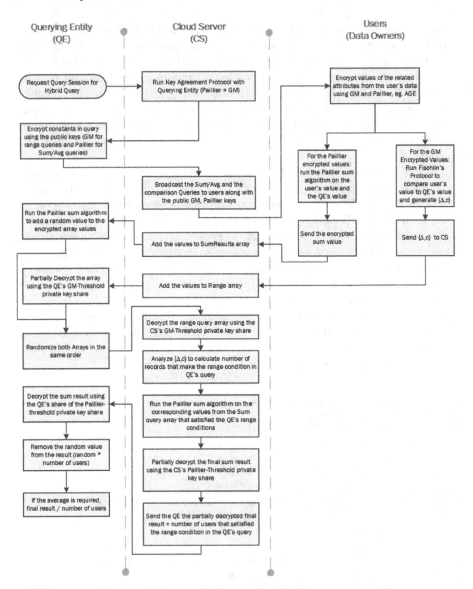

Fig. 2. Hybrid query execution flowchart

5   Hiding Query Attributes

In this section we will address the threat we described in Sect. 2.4. In the solution we presented so far, the query attributes are sent to the clients in plaintext. The purpose of keeping them in plaintext was to give the control to the data owners to decide which attributes can be queried. However, hiding the query attributes will further preserve the privacy of the QE without decreasing the privacy of the data owners.

We now present an extension to the protocol to also hides the query attributes from the DOs. To hide the query attributes, we modify the query distribution phase and the data owner query execution phase.

5.1 Modified Query Distribution Phase

To hide the attributes from the DOs we will first send a query request to the DOs. Each DO then generates a Paillier key pair (DO_pk_{pa} and DO_PK_{pa}) and uses the public key to encrypt all their data and send it to the CS in an array along with the generated public key. The CS would then be able to choose the attributes related to the QE's query from the array sent by the DO. The CS then creates a response array with only the encrypted values of the attributes in the query.

The CS then chooses a random value for each attribute. For example, *if the QE query was concerned with the DOs' "age" and "salary", we would have a one random value for "age" and another for "salary". The same random value would be used for the "age" attribute for all DOs.*

The CS then uses the generated public Paillier key to encrypt the random values and then add them to the encrypted values in the response array it created. The response array is then sent to the DO.

Phase 2. Hybrid queries

1: $QE \rightarrow CS : Q$
2: $CS :$
3: *for each* Q_a *in* Q *do* {
4: $Rand[i] = GenerateRandomValue();$}
5: $CS \rightarrow DOs : QueryRequest()$
6: *Each DO :*
7: $GeneratePaillierSessionKey()$ % Generates Public S_pk_{pa} and Private S_PK_{pa} Keys
8: *for each* Q_a *in* $DOdata$ *do* {
9:     $DATA_{pa}[i] = Enc_{S\_pk_{pa}}(DOdata.Q_{a_i}))$ }
10: $DOs \rightarrow CS : DATA_{pa}[\,]; S\_pk_{pa}$
11: $CS :$
12: Extract the elements from $DOdata[\,]$ that correspond to the Q_as in Q
13: *for each* $DOdata.Q_a$ *in* Q_a *do* {
14:     $attribute[i] = PaillierSum(DATA_{pa}[i], Enc_{S\_pk_{pa}}(Rand[i]))$
15:     if Q_a is related to Sum/Avg part of the query {
16:         $cryptosystem[i] = Paillier$}
17:     else if Q_a is related to the comparison part of the query {
18:         $cryptosystem[i] = GM$}
19: }
20: $CS \rightarrow DO : attribute[\,] ; cryptosystem[\,]; pk_{pa}; pk_{gm} ; GMEnc_{pk_{gm}}(Q_c[\,])$

5.2 Modified Data Owner Query Execution Phase

In this phase, the data owners are to execute the query without knowing which of their attributes are being queried. The participation of the DOs is needed in this phase because it is not possible to run the encrypted homomorphic operation without having the data encrypted with the session public key. For that reason, the DO needs to decrypt the values in the *attribute*[] array to be able to execute the homomorphic operations on the QE's query.

For the sum/avg part of the query, the DO simply re-encrypts the value of the attribute using the public Paillier session key pk_{pa} and add the result to $results_{pa}[]$. For the comparison part of the query, the DO needs to compare the re-encrypted value to the GM.

Phase 3. Hybrid queries Sum/Avg queries

1: *for each attribute*[] *do* {
2: *if* $(cryptosystem[i] == Paillier)$ {
3:     $results_{pa}[counterPA] = Enc_{pk_{pa}}(PaillierDecrypt(attribute[i], S\_PK_{pa})$
4:     $counterPA + +$ }
5: *else if* $cryptosystem[i] == GM$ {
6:     $results_{gm}[counterGM] = RunFischlin(GMEnc_{pk_{gm}}(Q_{c_{counterGM}}),$
       $GMEnc_{pk_{gm}}(PaillierDecrypt(attribute[i], S\_PK_{pa}))$
7:     $counterGM + +$ }}
8: $DOs \rightarrow CS : results_{pa}[]; results_{gm}[]$

---

6 Differentially Private Query Results

In this section, we will be addressing the threat model described in Sect. 2.4. The mechanism we presented in Sect. 4 securely executes aggregate and comparison queries while preserving the privacy of the QE and the DOs. The mechanism produces accurate results for each query. However, if the QE were to create a query with the goal of exposing PII about a specific DO, then the identity and data of a specific DO could be revealed.

To address this issue, we present an extension to the mechanism proposed in Sect. 4 to add differential privacy. Rather than returning the accurate result, the result becomes an approximation of the actual result. Formally, differential privacy is defined as follows:

A randomized function K gives ϵ-differential privacy if for all data sets D and D' differing on at most one row, and all $S \subseteq Range(K)$,
$$Pr[K(D) \in S] \leq exp(\epsilon) \times Pr[K(D') \in S]$$

Differential privacy ensures that the risk to a DO's privacy should not substantially decrease when responding to statistical queries. This means the knowledge a curious QE can gain about a specific DO, is not affected by the participation of the DO in responding to the query. This provides DOs with the assurance that the risk to the DO's privacy for participating and responding to a query is low. One method for adding noise involves the use of the Laplace mechanism to add random noise that conforms to the Laplace statistical distribution [17].

The Laplace distribution has two parameters, the location and scale parameters. The value of the location parameter will be set to zero to keep the noisy result close to the accurate result. The scale parameter is directly proportional to its standard deviation, or noisiness. The value we choose as the scale depends on the value of the privacy parameter ϵ, and the QE's query. Moreover, it corresponds to the maximum difference a DO can have on the result of the query f. This is known as the sensitivity of the query f, and can be defined mathematically as follows:

$$\triangle f =_{D,D'}^{max} \|f(D) - f(D')\|_1$$
$$=_{D,D'}^{max} \sum_{i=1}^{d} |f(D)_i - f(D')_i|$$

for all D, D' databases differing in at most one row. Dwork *et al.* [18] presents a proof showing that by adding a random Laplace $(\triangle f / \epsilon)$ variable to a query, ϵ-differential privacy is guaranteed.

6.1 Query Sensitivity

In the case of count queries, the maximum effect a DO can have on a query result is 1, for that reason the query sensitivity would always be equal to 1. For sum queries, the query sensitivity would be equal to the maximum value an attribute could have. In our mechanism, the CS would have a list of all attributes and the maximum value an attribute can have. For that reason, the query sensitivity would also be equal to 1 for binary value attributes. For larger value attributes, the sensitivity would be equal to the maximum value of that attribute. For example, if we wanted to execute a sum query on the "Diabetes" attribute, which specifies whether a DO has diabetes, then the sensitivity would be equal to 1. If we wanted to execute a sum query on the "Salary" attribute, it would be equal to the maximum defined salary value. Finally, for average queries the value of the sensitivity is equal to the maximum value defined value for the queried attribute divided by the number of DOs responding to the query.

6.2 Modified Protocol

To modify the protocol we introduced in Sect. 4, we first need to modify the noisy query response before returning it to the QE. There are two modifications that need to be made to the noise value. First of all, for the noisy query response to make sense, the value must be an integer, and for that reason we will be rounding the laplace noise before adding it to the actual result.

The second modification is to ensure the noisy query response is not a negative value. The laplace noise can be positive or negative, but for the query response to make sense we need the noisy response to be zero or more. In our mechanism, the query is executed by the CS, which means that for comparison queries, the CS would know the number of entities that satisfy the query conditions. We do not consider this a security concern because the CS will not be able to know what the query constants are or which DOs satisfy the query conditions. Since the CS knows the noise value and the number of DO that satisfy the query conditions, it would also know whether the noisy response is positive or negative. In the case of a negative noisy query response, the CS would always respond with a zero. However, in the case of sum or average queries, we propose aggregating the query sensitivity to the noisy response. This would enable the QE to decrypt the noisy query response. If the result is negative, the QE would consider the result to be zero. The modifications to the protocol would only change the cloud server execution phase we present in Sect. 4.7:

Phase 4. Hybrid queries

1: $CS \rightarrow QE : results_{gm}[\,]; results_{pa}[\,]$
2: $QE : PaillierSum(rand; results_{pa}[\,])$
3: $QE : Randomize(results_{gm}[\,]; results_{pa}[\,])$
4: $QE : PartiallyDecGMResult[\,] =$
    $GMDecrypt(results_{gm}[\,]; PK_{gm_1})$
5: $QE \rightarrow CS : PartiallyDecGMResult[\,]; results_{pa}$
6: $CS : DecryptedResult[\,] =$
    $GMDecrypt(PartiallyDecGMResult[\,]; PK_{gm_2})$
7: $CS : if(FischlinResAnalysis(DecryptedResult[i]))$
8:     $\{sumResult += results_{pa}[i];$
9:     $counter + +; \}$
10: $sumResult = PaillierSum(sumResult, LaplaceNoise(0, querySensitivity(Q)))$
11: $sumResult = PaillierSum(sumResult, querySensitivity(Q))$
12: $CS : PartiallyDecResult =$
    $PaillierDecrypt(sumResult; PK_{pa_2})$
13: $CS \rightarrow QE : PartiallyDecResult; counter; querySensitiviy(Q)$
14: $QE : FinalResult = PaillierDecrypt(PartiallyDecResult; PK_{pa_1}) - (rand *$
    $counter) - querySensitiviy(Q)$
    If average is needed:
15: $QE : Avg = FinalResult/counter$

---

7   Security Analysis

In this section, we address the security threats mentioned in Sect. 2.4:

Approach. Our solution aims at protecting the confidentiality of the querying entity data by preventing the CS from accessing the private keys. The querying entity encrypts the query constants with the session public key. The CS and the data owners can encrypt data and run homomorphic computations on the data using only public keys. The querying entity is also unable to access the final result without having the private key. However, the CS needs the private keys to be able to preform comparisons on encrypted data. To enable the CS to complete these computations, we rely on threshold encryption with a randomization algorithm. Together, they allow the CS to complete the secure computations with the assistance of the QE while protecting the confidentiality of the data owners' data and the QE's query constants.

Guarantees. Confidentiality is provided by our solution for query constants, DOs data, and the secure computation results. The solution does not hide the attributes in the query or the number of data owners responding to a query. However, the solution does protect the identities of the data owners that satisfy the query range and prevents the CS from correlating a any results to specific DOs. The security of our solution is not perfect: The number of DOs that satisfy the range conditions in the query is revealed to the CS. Also, the number of DOs that responded to a query is revealed to the QE, but the number of those that satisfy the range conditions is not revealed. Finally, the solution does not protect against queries that target specific data owners. For example, a QE can send a query to find if a specific person has specific disease. By sending a sum query of all DOs that have that disease and combining the query with other identifiers such as location, age, gender, etc. The result of the query can reveal the data of a specific DO. To minimize this problem, the solution prevents the CS from sending the final result of the query to the QE in case the result of the query condition only involves a single data owner. This technique is not prefect because a malicious QE can craft complex queries to reveal data of specific DOs, and this attack is not addressed by our solution. More intuitively, our solution provides the following security properties:

– *The query constants and the random values cannot be recovered without the participation of the QE:* The QE sends the sanitized query constants and the random values encrypted using a semantically secure cryptosystem: Paillier-threshold in the case of sum queries, or GM-threshold in the case of range queries. This means that the query constants and the random values cannot be recovered without the participation of the QE [9,10].
– *The cloud server cannot correlate any of the results to specific data owners:* In the *cloud server query execution phase*, the CS sends the results array $results_{gm}$ to the QE for partial decryption in the case of range queries, and sends both arrays $results_{gm}$ and $results_{pa}$ in the case of a hybrid query. Randomizing the order of $results_{gm}$ along with the partial decryption prevents the CS from correlating a result to a specific DO. However, in the case of hybrid queries, the QE has to randomize both arrays in the same order before sending them back to the CS. This renders the partial decryption ineffective in preventing the CS from correlating a result to a specific DO. It is for this reason that we must add a random value to all the elements in the $results_{pa}$ array using the Paillier sum algorithm. This ensures that the CS

executes the query using the randomized array $results_{pa}$ and the partially decrypted $results_{gm}$ without being able to correlate any of the results to a specific DO.

- **Data owner data cannot be decrypted or calculated by the cloud server or the querying entity:** In regard to range queries, the DOs do not send their actual data encrypted, but instead send the results of the Fischlin's private comparison protocol (Δ, c), which are then used to compare the encrypted values.

  In the case of sum/avg or hybrid queries, the DOs send their data encrypted using the Paillier-threshold cryptosystem after adding a random value to it using the Paillier sum algorithm. Adding the random value prevents the CS from calculating the DO data in case the CS sends the QE the $results_{pa}$ array in place of the $results_{gm}$ array in the hybrid case of the *cloud server query execution phase*, thus receiving a partially decrypted $results_{pa}$ array, which it can then decrypt, using its Paillier-threshold private key share, to access the actual DO data. Under our semi-honest adversary model assumption, we do not need to add this extra step; however, due to the simplicity of this attack and due to the fact that it would expose the DOs' actual data, we felt that adding this step is a necessity.

- **Data owners cannot know what attributes are being queried by the querying entity:** In the extension we present in Sect. 5, the DO send their data encrypted using a public key, which they generated for that session. No entity has access to the corresponding private key, which means the confidentiality of the DO's data is protected. The DO also sends the public keys to allow the CS to perform the homomorphic addition of random values to the DO's values corresponding to the attributes related to the QE's query. The CS then sends the noisy values back to the DO in random order. This ensures that the DO will receive noisy values in random order, thus preventing the DO from knowing which attributes the QE was interested in.

- **The QE cannot learn about Personally Identifiable Information (PII) about a specific data owner:** In the extension we presented in Sect. 6, the CS adds a noisy value to the final response to protect the privacy of DOs in the system. The noise is chosen according to the type of query to ensure that the response is as accurate as possible while satisfying the definition of differential privacy. The noise is chosen from a laplace distribution where the maximum amount of noise is equal to the maximum difference a single DO can have on the response of query. This ensures that even if the QE manages to create a query that should result in PII, it cannot be sure whether the final result of the query is PII or noise.

8   Performance Evaluation

In this section, we present our experimental result. We rely on the open-source projects used in [19]. The code is written in Java and executed on Amazon's elastic cloud (EC2) [20]. Amazon allows users to run their applications on virtual machines, called instances. Amazon offers a variety of instance types with different configurations of CPU, RAM memory, and ROM storage. Amazon EC2 is built on commodity hardware, and over time there may be several different types of physical hardware underlying EC2 instances. Using this method allows Amazon to provide consistent amounts of processing power regardless of the actual underlying physical hardware. A single EC2 compute unit produces the equivalent CPU capacity of 1.0–1.2 GHz 2007 Opteron or 2007 Xeon processor. We run our implementation on an Windows instance with the following specs:

- 128 vCPUs
- 1,952 GiB of memory
- 2 × 19, 20 GB of SSD instance storage
- Instance Type: x1.32xlarge

The execution time in our work depends greatly on the type of query and the number of attributes in the query. To demonstrate the required processing time, we take sum and hybrid queries as an examples. Sum queries simply aggregate a value of a common attribute among data owners. Hybrid queries are aggregate queries that also include comparisons. For that reason, they are the most expensive ones since they utilizes all the building blocks. As previously mentioned, our goal is to allow a QE to query privately encrypted distributed data with the assistance of a semi-trusted cloud. We run our code assuming 100, 1000, 10 k, and 100 k data owners to demonstrate the practicality of our approach. We do not take network delays into account. We also assume data owners run their computations in parallel. To test the performance we execute the following two queries:

$$AVG \ (doctorVisitsPerYear) \ FROM \ data\_owners$$
$$AVG \ (doctorVisitsPerYear) \ FROM \ data\_owners \ WHERE \ age \ > \ 30 \ AND \ diabetes \ = \ 1$$

Our implementation shows how the overall load on the CS increases with the number of data owners as shown in Figs. 3 and 4. Our results show that executing the sum query on 10 k DOs requires 0.023 seconds. However, hybrid queries require 2.34 min due to the high computational overhead of Fischlin's protocol. These computations can be done in parallel, which allows us to reduce the computation time with the use of cluster computing solutions such as Apache Spark [21]. In fact, with the assistance of 100 clusters of the same instance used, the hybrid query can be executed for 1 million data owners in less that 3 min. These results show that our approach can be reasonably used for large set of data owners to provide the results in timely manner.

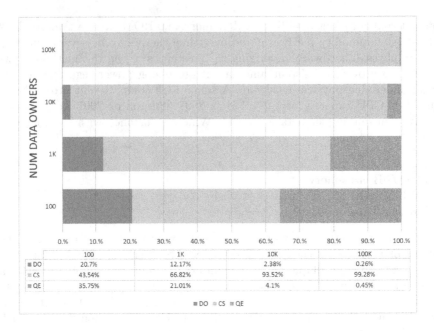

Fig. 3. The processing load distribution for sum queries

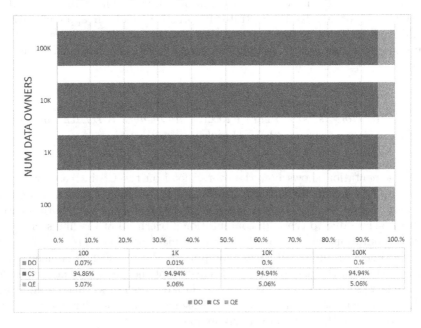

Fig. 4. The processing load distribution for hybrid queries

9   Related Work

In this section, we will discuss private database outsourcing, which deals with problem of hiding data from untrusted service providers. Most existing approaches rely on encryption to preserve the privacy of the data stored on the cloud [22]. However, relying on encryption to secure the data causes new problems, such as querying the encrypted data. Previous work in this domain is divided into two categories, one that relies on a trusted server to assist in executing the query [23], and another that relies on the assistance of a semi-trusted server [24–28].

The use of a trapdoor encryption method was suggested by Song et al. [24], in which two layers of encryption were used. The first layer uses a symmetric key with a secret key, while the second uses a pseudo-random number generator and two pseudo-random functions. However, their method only allowed querying for equality. The addition of secure indexes in each tuple was suggested by Hacigumus et al. [25,26]. A similar solution was also proposed by Damiani et al. [27]. To create the indexes in their methods, the data is classified by a collusion hash function thus preventing the data from being classified sequentially. A method allowing range queries was suggested by Agrawal et al. [28]. In their work, an encryption protocol was proposed allowing for the comparisons to be executed on the encrypted data directly. It was also assumed in their work that the order of the data was confidential. Boneh et al. [29] proposed a solution based on Hidden Vector Encryption (HVE). The solution allows queries on encrypted data to produce tokens for testing supported query predicates. Tokens allow users to test the predicate of a ciphertext without leaking any information on the plaintext. This allows for comparison and subset queries.

Although the described techniques have been proposed to allow for querying encrypted data stored on a CS, they cannot be adopted to solve the problem in this paper for several reasons. First, to allow the CS to evaluate the query on the encrypted data, the QE must send the query to the CS after encrypting it using the same cryptosystem and key used by the DOs. The query may then be sent by the CS to the DOs, where the query can be decrypted. Second, a common approach in the techniques above is to send a set of encrypted records to the QE for filtration and further processing [25,30]. Therefore, the CS would be revealing more information than the final query result to the QE.

CryptDB [6] has been proposed to execute SQL queries over encrypted data. It depends on a trusted proxy that maintains all the secret and public keys and transforms the users' SQL queries to a query that can be executed over encrypted records. CryptDB has low overhead on query execution time; however, it requires pre-processing user data and storing it encrypted in a central location. It also requires a trusted proxy which can compromise the data of all logged in users.

10   Conclusion

In this paper, we described the problem and challenges of securely querying privately encrypted data, where data owners store their data in the cloud encrypted with their own private keys, while preserving the privacy of all the parties involved. We proposed a protocol that allows third parties to execute various types of queries on privately encrypted data stored in an untrusted cloud while preserving the privacy of both; the data owners data and identity, as well as the querying entity's query attributes and constants. The protocol relies on semantically secure probabilistic cryptosystems and differential privacy. It also allows for range, sum/avg, and hybrid queries. The protocol relies on two cryptosystems along with Fischlin's protocol for private comparisons. We also present the results of our implementation, which show the feasibility and scalability of our approach. Finally, we analyzed the security of our protocol.

For future work, we will be looking into methods of ensuring the preservation of data owner privacy by analyzing current and previous queries. This would prevent third parties from inferring additional information about specific data owners.

References

1. Microsoft health vault. http://www.healthvault.com. Accessed Oct 2017
2. Li, M., Yu, S., Ren, K., Lou, W.: Securing personal health records in cloud computing: patient-centric and fine-grained data access control in multi-owner settings. In: Jajodia, S., Zhou, J. (eds.) SecureComm 2010. LNICSSITE, vol. 50, pp. 89–106. Springer, Heidelberg (2010). https://doi.org/10.1007/978-3-642-16161-2_6
3. Akinyele, J.A., Pagano, M.W., Green, M.D., Lehmann, C.U., Peterson, Z.N.J., Rubin, A.D.: Securing electronic medical records using attribute-based encryption on mobile devices. In: Proceedings of the Workshop on Security and Privacy in Smartphones and Mobile Devices. SPSM 2011, pp. 75–86. ACM (2011)
4. Narayan, S., Gagné, M., Safavi-Naini, R.: Privacy preserving EHR system using attribute-based infrastructure. In: Proceedings of the Cloud Computing Security Workshop. CCSW 2010, pp. 47–52, New York. ACM (2010)
5. PwC: Putting data security on the top table: how healthcare organisations can manage information more safely, June 2013
6. Popa, R.A., Redfield, C.M.S., Zeldovich, N., Balakrishnan, H.: CryptDB: protecting confidentiality with encrypted query processing. In: Proceedings of the ACM Symposium on Operating Systems Principles, pp. 85–100 (2011)
7. Tu, S., Kaashoek, M.F., Madden, S., Zeldovich, N.: Processing analytical queries over encrypted data. In: Proceedings of the VLDB Endowment, vol. 6, pp. 289–300. VLDB Endowment (2013)
8. Aljumah, F., Pourzandi, M., Debbabi, M.: Privacy-preserving querying mechanism on privately encrypted personal health records. In: International Conference on Informatics, Health & Technology (ICIHT), pp. 1–8. IEEE (2017)
9. Paillier, P.: Public-key cryptosystems based on composite degree residuosity classes. In: Stern, J. (ed.) EUROCRYPT 1999. LNCS, vol. 1592, pp. 223–238. Springer, Heidelberg (1999). https://doi.org/10.1007/3-540-48910-X_16
10. Goldwasser, S., Micali, S.: Probabilistic encryption & how to play mental poker keeping secret all partial information. In: Proceedings of the 14th Annual ACM Symposium on Theory of Computing (1982)
11. Katz, J., Yung, M.: Threshold cryptosystems based on factoring. In: Zheng, Y. (ed.) ASIACRYPT 2002. LNCS, vol. 2501, pp. 192–205. Springer, Heidelberg (2002). https://doi.org/10.1007/3-540-36178-2_12
12. Barnett, A., Smart, N.P.: Mental poker revisited. In: Paterson, K.G. (ed.) Cryptography and Coding 2003. LNCS, vol. 2898, pp. 370–383. Springer, Heidelberg (2003). https://doi.org/10.1007/978-3-540-40974-8_29

13. Boneh, D., Franklin, M.: Efficient generation of shared RSA keys. J. ACM **48**(4), 702–722 (2001)
14. Yao, A.C.: Protocols for secure computations. In: Proceedings of the 23rd Annual Symposium on Foundations of Computer Science. SFCS 1982, pp. 160–164. IEEE Computer Society (1982)
15. Fischlin, M.: A cost-effective pay-per-multiplication comparison method for millionaires. In: Naccache, D. (ed.) CT-RSA 2001. LNCS, vol. 2020, pp. 457–471. Springer, Heidelberg (2001). https://doi.org/10.1007/3-540-45353-9_33
16. Shamir, A.: How to share a secret. Commun. ACM **22**(11), 612–613 (1979)
17. Dwork, C., McSherry, F., Nissim, K., Smith, A.: Calibrating noise to sensitivity in private data analysis. In: Halevi, S., Rabin, T. (eds.) TCC 2006. LNCS, vol. 3876, pp. 265–284. Springer, Heidelberg (2006). https://doi.org/10.1007/11681878_14
18. Dwork, C.: A firm foundation for private data analysis. Commun. ACM **54**(1), 86–95 (2011)
19. Barouti, S., Aljumah, F., Alhadidi, D., Debbabi, M.: Secure and privacy-preserving querying of personal health records in the cloud. In: Atluri, V., Pernul, G. (eds.) DBSec 2014. LNCS, vol. 8566, pp. 82–97. Springer, Heidelberg (2014). https://doi.org/10.1007/978-3-662-43936-4_6
20. Amazon Web Services: Amazon elastic compute cloud (EC2) documentation. http://aws.amazon.com/documentation/ec2/. Accessed Oct 2017
21. Apache Software Foundation: Apache spark. http://spark.apache.org/. Accessed Oct 2017
22. Sion, R.: Secure data outsourcing. In: Proceedings of the Conference on Very Large Data Bases. VLDB 2007, pp. 1431–1432 (2007)
23. Iyer, B., Mehrotra, S., Mykletun, E., Tsudik, G., Wu, Y.: A framework for efficient storage security in RDBMS. In: Bertino, E., Christodoulakis, S., Plexousakis, D., Christophides, V., Koubarakis, M., Böhm, K., Ferrari, E. (eds.) EDBT 2004. LNCS, vol. 2992, pp. 147–164. Springer, Heidelberg (2004). https://doi.org/10.1007/978-3-540-24741-8_10
24. Song, D., Wagner, D., Perrig, A.: Practical techniques for searches on encrypted data. In: IEEE Symposium on Security and Privacy (2000)
25. Hacıgümüş, H., Iyer, B., Li, C., Mehrotra, S.: Executing SQL over encrypted data in the database-service-provider model. In: Proceedings of the 2002 ACM SIGMOD International Conference on Management of Data (2002)
26. Hacıgümüş, H., Iyer, B., Mehrotra, S.: Efficient execution of aggregation queries over encrypted relational databases. In: Lee, Y.J., Li, J., Whang, K.-Y., Lee, D. (eds.) DASFAA 2004. LNCS, vol. 2973, pp. 125–136. Springer, Heidelberg (2004). https://doi.org/10.1007/978-3-540-24571-1_10
27. Damiani, E., Vimercati, S., Jajodia, S., Paraboschi, S., Samarati, P.: Balancing confidentiality and efficiency in untrusted relational DBMSs. In: Proceedings of the 10th ACM Conference on Computer and Communications Security (2003)
28. Agrawal, R., Kiernan, J., Srikant, R., Xu, Y.: Order preserving encryption for numeric data. In: ACM SIGMOD International Conference on Management of Data (2004)
29. Boneh, D., Waters, B.: Conjunctive, subset, and range queries on encrypted data. In: Vadhan, S.P. (ed.) TCC 2007. LNCS, vol. 4392, pp. 535–554. Springer, Heidelberg (2007). https://doi.org/10.1007/978-3-540-70936-7_29
30. Hore, B., Mehrotra, S., Tsudik, G.: A privacy-preserving index for range queries. In: Proceedings of the International Conference on Very Large Data Bases. VLDB 2004, vol. 30, pp. 720–731. VLDB Endowment (2004)

Comparison of Adaptive Neuro-Fuzzy Inference System (ANFIS) and Gaussian Process for Machine Learning (GPML) Algorithms for the Prediction of Norovirus Concentration in Drinking Water Supply

Hadi Mohammed[1(✉)], Ibrahim A. Hameed[2], and Razak Seidu[1]

[1] Water and Environmental Engineering Group, Department of Marine Operations and Civil Engineering, Faculty of Engineering, Norwegian University of Science and Technology (NTNU), Postboks 1517, 6025 Ålesund, Norway
{hadi.mohammed,rase}@ntnu.no
[2] Department of ICT and Natural Sciences, Faculty of Information Technology and Electrical Engineering, Norwegian University of Science and Technology (NTNU), Postboks 1517, 6025 Ålesund, Norway
ibib@ntnu.no

Abstract. Monitoring of Norovirus in drinking water supply is a complicated, rather expensive, process. Norovirus represent a leading cause of acute gastro-enteritis in most developed countries. Modeling of general microbial occurrence in drinking water is a very active field of study and provides reliable information for predicting microbial risks in drinking water. In this work, adaptive neuro-fuzzy inference system (ANFIS) and Gaussian Process for Machine Learning (GPML) are proposed as predicting models for the total number of Norovirus in raw surface water in terms of water quality parameters such as water pH, turbidity, conductivity, temperature and rain. The predictive models were based on data from Nødre Romrike Vannverk water treatment plant in Oslo, Norway. Based on the model performance indices used in this study, the GPML model showed comparable accuracy to the ANFIS model. However, the ANFIS model generally demonstrated more superior prediction ability of the number of Norovirus in drinking water, with lower MSE and MAE values relative to the GPML model. In addition, the ability of the ANFIS model to explain potential effects of inter-actions among the water quality variables on the number of Norovirus in the raw water makes the technique more efficient for use in water quality modeling.

Keywords: GPML · ANFIS · Norovirus · Water pH · Turbidity · Water conductivity · Temperature · Conductivity · Rain

1 Introduction

Norovirus is increasingly recognized as a leading cause of non-bacterial gastrointestinal infections, and is a major cause of waterborne disease outbreaks worldwide [1, 2] an estimated economic burden of $4.2 billion in direct health system costs worldwide [3]. Moreover, enteric Noroviruses have very low infectious dose and like other viruses,

© Springer-Verlag GmbH Germany 2017
A. Hameurlain et al. (Eds.): TLDKS XXXV, LNCS 10680, pp. 74–95, 2017.
https://doi.org/10.1007/978-3-662-56121-8_4

their small size, low inactivation rates and the inability to culture them make their removal difficult and their susceptibility to disinfection largely remain unknown [4–6]. Further, results of some studies suggest that the viruses may be resistant to wastewater treatment since effluents are not completely devoid of the them [7–9], resulting in source water contamination when effluents of wastewater treatment plants are discharged to surface waterbodies [10–12].

Although recent molecular methods have improved the detection, identification and characterization of Norovirus in the environment and clinical samples, widespread emergence of the virus still present a challenge to the detection technique [13]. These complicate the health risks associated with the use of drinking and recreational water from contaminated water resources. Mitigating the morbidity and mortality associated with waterborne infections of Norovirus accordingly require proactive measures to augment rather costly monitoring exercises that assess microbial quality of drinking water sources. The dynamics of Norovirus occurrence in raw water sources depends on a complex interaction of different variables, including environmental factors (e.g. temperature, rainfall etc.) [14]. There is very little understanding of the environmental factors that significantly trigger the occurrence of Norovirus [15]. Environmental factors such as rainfall and temperature are associated with increased concentrations of indicator pathogen in surface water and are noted as potential predictors of increased source water pathogen concentration [16, 18]. As with many water treatment plants worldwide, Norwegian water treatment plants (WTPs) do not monitor raw water sources for specific pathogenic organisms (including Norovirus) due to cost considerations.

Limitations regarding the lack of source water microbial quality data needed for microbial risk assessment have necessitated the use of mathematical models to predict the occurrence of pathogenic organisms in raw water sources. Predictive models, based on environmental and water quality parameters have been widely applied to improve the accuracy of raw water quality assessments, in order to assist watershed managers in making informed decisions regarding the protection of public health. Physically based techniques such as hydrodynamic models have been used to monitor microorganism generation, fate and transport in surface water sources [19, 20].

Whereas data-driven techniques such as regression analysis have widely been used, other artificial intelligence (AI) techniques such as artificial neural network and adaptive neuro-fuzzy inference system (ANFIS) are recently being applied in predicting the concentration of microbial organisms in water sources [19–23]. In a recent study in Norway, Peterson et al. (2016) applied a quantitative microbial risk approach to model the concentration of Norovirus in surface water based on *E. coli* and *C. perfringens* concentrations with assumptions regarding the source of fecal contamination [19]. Although poor correlations were found between the pathogen and indicator data, the approach provided an insight into potential level of Norovirus contamination in a typical surface water body in Norway [19]. In this paper, Adaptive Neuro-Fuzzy Inference System (ANFIS) is built to predict the concentration of Norovirus in the raw water source of the Nødre Romrike Water Treatment Plant in Oslo, based on measured rainfall in the catchment of the water supply system and water quality parameters such as water temperature, turbidity, conductivity and pH. In addition, Gaussian Processes for Machine Learning (GPML) modeling approach is used on the same dataset to predict the count of Norovirus in the raw water. The performances of the two modeling approaches are then compared using mean square prediction errors, mean absolute errors and the coefficient of multiple determination, R^2 values. The paper is organized as follows: ANFIS model is presented in Sect. 2. In Sect. 3, the data and modelling approaches are presented. Results are presented in Sect. 4. In Sect. 5, concluding remarks are drawn and suggestions for future work are presented.

2 Methodology

2.1 Adaptive Neuro-Fuzzy Inference System (ANFIS)

ANFIS is a well-known artificial intelligence technique that has been used in hydrological processes [22]. With respect to water quality monitoring, the technique has been widely used to model treatment processes, estimation of concentrations of disinfection byproducts as well as other water quality indices of groundwater [21, 23–25]. By analyzing mapping relationships between input and output data, ANFIS optimizes the distribution of membership functions by using a hybrid learning algorithm consists of a combination of least-squares and back-propagation gradient descent algorithm [26]. In this paper two membership functions are assigned for each input variable and therefore the ANFIS model will generate 64 rules (i.e., 26 rules). The proposed ANFIS (Fig. 5 has six inputs; pH, turbidity, conductivity, rain, temperature and seasonality and one output, the concentration of Norovirus. Each input is represented by two fuzzy sets, and the output by a first-order polynomial of the inputs. The ANFIS extracts n rules mapping the inputs to the output from the input/output dataset. A typical Sugeno-fuzzy rule can be expressed in the following form:

$$R_i: IF\ pH\ is\ A_{1,j}$$
$$AND\ Turbidity\ is\ A_{1,j}$$
$$\vdots \tag{1}$$
$$AND\ Temperature\ is\ A_{m,j}$$
$$THEN\ NoV_i = f_i(pH,\ Turbidity,\ \ldots,\ Temperature)$$

Where $A_{1,j}$, $A_{2,j}A_{1,j}$, $A_{2,j}$, $\ldots$, $A_{m,j}$ are fuzzy sets or fuzzy labels used to fuzzify each input, NoV_i (i.e., Norovirus count of rule i) is either a constant or a linear function of the input variables of the model. When NoV_i is constant, a zero-order Sugeno fuzzy model is obtained in which the consequent of a rule is specified by a singleton. When NoV_i is a first-order polynomial of the inputs, the consequent of a rule is a polynomial that takes the form:

$$NoV_i = k_{i0} + k_{i1}pH + k_{i2}Turbidity + \ldots + k_{im}Temperature \tag{2}$$

A   first-order   Sugeno   fuzzy   model   is   obtained   where k_{i0}, k_{i0}, k_{i1}, k_{i2}, $\ldots$ and $k_{im}k_{i1}$, k_{i2}, $\ldots$ and k_{im} are a set of parameters specified for rule i [26]. An ANFIS model, as shown in Fig. 1, is normally represented by a six-layer feed-forward neural network representing the architecture of a first-order Sugeno fuzzy model

[27]. The first layer is called input layer. Neurons in this layer simply pass external crisp signals to the second layer. The second layer is called the fuzzification layer. Neurons in this layer perform fuzzification. Fuzzification neurons have a bell-shaped activation function specified as:

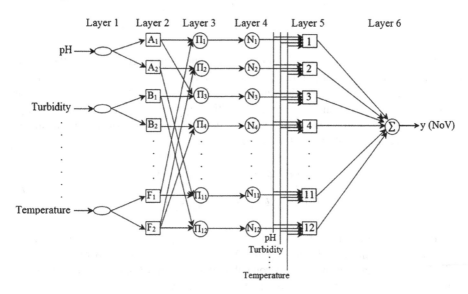

Fig. 1. ANFIS architecture for first order Sugeno fuzzy model as used in this study.

$$y_i^{(2)} = \frac{1}{1 + \left(\dfrac{x_i^{(2)} - a_i}{c_i} \right)^{2b_i}} \tag{3}$$

where x_i is the input and y_i is the output of layer 2; a_i, b_i and c_i are the parameters that control, respectively, the center, width and slope of the bell activation function of neuron i. The third layer is called the rule layer. Each neuron in this layer corresponds to a single Sugeno-type fuzzy rule, as it is shown by Eq. (1). A rule neuron receives inputs from the respective fuzzification neurons and calculates the firing strength of the rule it represents. In an ANFIS, the product operator is used to evaluate the conjunction of the rule antecedents:

$$y_i^{(3)} = \prod_{j=1}^{k} x_{ij}^{(3)} = \mu_{i1} \times \mu_{i2} \times \cdots \times \mu_{ik} \tag{4}$$

where x_{ij} is the input from neuron j in layer 2 to neuron i in layer 3, and y_i is the output of layer 3.

Layer 4 is called the *normalization layer*. Each neuron in this layer receives inputs from all neurons in the rule layer, and calculates the normalized firing strength of a given rule. The normalized firing strength is the ratio of the firing strength of a given rule to the sum of the firing strengths of all rules. It represents the contribution of a given rule to the final result. The output of neuron i in layer 4 is obtained as:

$$y_i^{(4)} = \frac{x_{ij}^{(4)}}{\sum_{j=1}^{n} x_{ij}^{(4)}} = \frac{\mu_i}{\sum_{j=1}^{n} \mu_j} = \bar{\mu}_i \tag{5}$$

where x_{ij} is the input from neuron j in layer 3 to neuron i in layer 4, and y_i is the output of layer 4, and n is the total number of fuzzy rules. Layer 5 is called the defuzzification layer. Each neuron in this layer is connected to the respected normalization neuron in the normalization layer, and also receives initial inputs; $x_1, x_2,..., x_m$ (pH, turbidity, conductivity, rain, winter season, and temperature). A defuzzification neuron calculates the weighted consequent value of a given rule as:

$$\begin{aligned} y_i^{(5)} &= x_i^{(5)} \left(k_{i0} + k_{i1}x_1 + \rightleftharpoons + k_{im}x_m \right) \\ &= \bar{\mu}_i \left(k_{i0} + k_{i1}x_1 + \rightleftharpoons + k_{im}x_m \right) \end{aligned} \tag{6}$$

where x_i is the input and y_i is the output of neuron i in layer 5, $k_{i0}, k_{i1}, ...,$ and k_{im} is a set of consequent parameters of rule i, defined by Eq. (2). Layer 6 is finally represented by a single summation neuron. This neuron calculates the sum of outputs of all defuzzification neurons and produces the overall ANFIS output y.

It is not necessary to have any prior knowledge of rule consequent parameters for an ANFIS to deal with the problem. The parameters of the consequent polynomials are initialized to zero values. ANFIS uses a hybrid-learning algorithm that combines the least-squares estimator and the gradient descent method to learn parameters of the consequent polynomials and to tune the parameters of the membership functions. The only prior information required from the user is the number of membership functions required to fuzzify each input variable. The universe of discourse of each input variable is divided equally between its respective membership functions to find its centers. The widths and slopes are set to allow sufficiently overlapping between the respective functions.

In the ANFIS training algorithm, each epoch is composed from a forward pass and a backward pass. In the forward pass, a training set of input patterns is applied to the ANFIS, neuron outputs are calculated on the layer-by-layer basis, and the least-squares estimator identifies rule consequent parameters. In the backward pass, the error signals are propagated back and the back-propagation algorithm is used to update/tune the membership functions parameters of the rule antecedents.

3 Gaussian Processes for Machine Learning (GPML)

Gaussian processes (GPs) [30] have convenient properties for many modeling tasks in machine learning and statistics. They can be used to specify distributions over functions without having to commit to a specific functional form. Applications range from regression over classification to reinforcement learning, spatial models, survival and other time series models. A GP is specified by a mean function and a covariance function. These functions are mostly difficult to specify fully a priori, and typically they are given in terms of hyperparameters, that is, parameters which have to be inferred. Any functions can be used as a mean and a covariance functions. The mean function is usually defined to be zero. Several covariance functions have been used in literature. However, a squared exponential (SE) is usually used as a predominant choice. Another source of difficulty is the likelihood function. For Gaussian likelihoods, inference is analytically tractable; however, in many tasks, Gaussian likelihoods are not appropriate, and approximate inference methods such as Expectation Propagation (EP), Laplace's approximation (LA) and variational bounds (VB) become necessary [31].

GPML, in this paper, is implemented using GPML toolbox [32]. The GPML toolbox provides a wide range of functionality for GP inference and prediction. It is designed to simplify the process of constructing GP models and make it easy to extend where a library of various mean, covariance and likelihood functions as well as inference methods is provided.

4 Modelling Approach

4.1 ANFIS Model Training

In this paper, various ANFIS models with various settings are used to predicted count of Norovirus in raw water in terms of a set of input variables: water pH, water turbidity, water conductivity, rain, water temperature, and finally seasonal effect (i.e., winter time). The model building process for the ANFIS consists of the following five steps: (1). Selection of the input and the output data for training ANFIS model (data set). (2). Normalization of the input and the output data attributes. (3). Training of the normalized data using a hybrid-learning algorithm; (4). Testing the goodness of fit of the model; and (5) Comparing the predicted output with the desired/target output. Each of these steps are presented as follows:

4.2 Dataset and Input – Output Selection for Model Training

In this paper, ANFIS and GPML models are developed to accurately predict the concentration of Norovirus in terms of precipitation and physico-chemical characteristics of the raw water source of the Nødre Romrike Vannverk (NRV) water treatment plant in Oslo, Norway. NRV is one of the largest water treatment plants in Norway supplying 42000 m^3 of drinking water to six municipalities [28]. The plant depends on the raw water from Glomma River, the largest river in Norway. Data used in the study are based on raw water samples at the intake of NRV in the period from January 2011 to April 2012, covering the four main seasons in Norway. Sampling and analysis of raw water for Norovirus (GI and GII) was conducted under an EU project (VISK) [28]. For a detailed description of the analysis, recovery and assessment of Norovirus concentration in the raw water interested readers are advised to refer to Grøndahl- Rosado et al. (2014) [29]. Data on precipitation was collected at a weather station located within the catchment of the raw- water intake (e-Klima) while physical and chemical parameters data of the raw water were drawn from NRV database. The main physical and chemical parameters accounted for were water temperature ($^{\circ}$C), water turbidity (NTU/mL), water conductivity (μS/cm), rainfall (mm/day), and water pH. A total of 156 data samples are used in this study.

4.3 Data Normalization

The input and the output data obtained are measured on different scales and therefore have to be normalized using mean and standard deviation to a notionally common scale. First the mean, x, and standard deviation, σ_x, of all the data variables individually were calculated. The values for each parameter were then normalized using the equation:

$$x_n = (x - \bar{x})/\sigma_x \qquad (7)$$

4.4 Training of Input Data

After obtaining the normalized data, the next step is to train the input data using proposed ANFIS. The normalized water quality variables, pH, turbidity, conductivity, rainfall, temperature and time are used as the predictors of count of Norovirus (NoV) in raw water. ANFIS model uses a hybrid-learning algorithm that combines the least- squares estimator and the gradient descent method to learn parameters of the consequent poly-nomials and to tune the parameters of the membership functions. The algorithm, by default, takes only 70% of the input data for training. So out of 156 samples only 109 are taken for training and these are selected randomly from the set of data. The remaining 47 samples are kept for validation and testing. Various experiments were conducted with the aim of achieving the optimal ANFIS model with the least MSE, simplest structure, and ability to estimate periods of low and high counts of Norovirus in the raw water, given the dataset available for this study. To achieve this objective, we conducted five different experiments with the dataset in the ANFIS model building process.

An initial model structure with all the six water quality variables used as inputs, with each assigned 2 membership functions 'low' and 'high'. This resulted in a total of fuzzy rules used to establish relationships between the water quality variables and the count of Norovirus in the raw water. The model was run with different epochs until the least mean square error (MSE) is achieved. In addition, a comparison between model predic-tions and observed counts of the virus is made to assess the ability of the model to predict periods of low and high counts of the virus in the raw water. After the initial configu-ration, the water quality variables that show significant correlations with the count of Norovirus (obtained from prior Pearson's correlation analysis) are used to fit the model, with two membership functions during training. Further, principal component analysis (PCA) are further applied to the input dataset to select water quality variables that distinctly affect the variation of Norovirus in the raw water. The variables obtained from the PCA are used to train the ANFIS model, also with two membership functions. Finally, to determine if any variations in the number of membership functions used in the training of the model will have any significant effect on the performance of the model, we once again used all the six water quality variables as inputs, and Norovirus as output, but with three membership functions (Generalized bell-shaped). The results of each of the five models are then compared in terms of their abilities to estimate high and low counts of the virus as well as the overall MSE values.

5   GPLM Model Development

The proposed GPML model has six inputs; water pH, turbidity, conductivity, rain, temperature and seasonality. The GPML toolbox in MATLAB is used [32] for constructing a GP predictive model for Norovirus using the aforementioned six descriptive features. GPs are used to formalize and update knowledge about distribution over functions. To set up a GP model, a mean function with an initial value of mean = 0 is chosen. A squared exponential covariance function with hyper-parameters $\psi = \{1, 2\}$ is used. A Gaussian likelihood function with hyperparameter (i.e., Gaussian noise with variance ρ) where $\rho = \{1\}$ is used with an expectation propagation (EP) approximate inference algorithm.

5.1   Comparisons of Actual Data and Predicted Data

After the testing is done, the ANFIS and the GPML models are saved. The mean absolute error (MAE) and mean- squared error (MSE) between actual and predicted outputs and the coefficient of determination, R^2 are used as performance indices of the models' accuracy. A graph is plotted between the actual output and the predicted output so that a comparison can be easily made.

6   Results

Figure 2 shows the distribution of the water quality parameters and their influence on Norovirus concentration. The measured Norovirus concentrations had large variations. While variations in the water pH and electrical conductivity remain low, the range of variations in measured values of water temperature, rainfall and turbidity remained high. For the months in which rainfall over the study area was high (mostly between July and December), the water temperature was continuously high. However, the turbidity level in the raw water reached its peak of 13 NTU in in the middle of June, just prior to the onset of elevated rainfall in mid-July. High Norovirus concentrations over the sixteen-month study period, the observed Norovirus concentrations were very high between January and April of 2011 (500 particles per liter), with intense variations. Subsequently however, few Norovirus particles are observed intermittently.

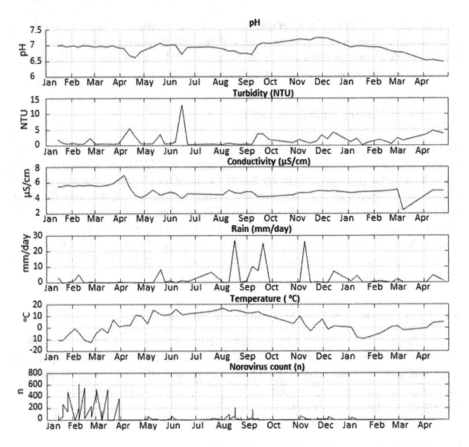

Fig. 2. Raw data from NRV from middle of January 2011 until the end of April 2012.

Figure 3 shows the box plot of the raw water quality parameters showing the medians, minimum, and maximum values of each individual variable. It is evident from this figure that significant outliers are present in the measured water quality variables. For the count of Norovirus in the raw water, the data is highly skewed, with majority of the observations constituting outliers. The correlation coefficient between the dependent (i.e., count of Norovirus) variable and independent variables (i.e., model inputs) are shown in Table 1. From the table, it can be concluded that the observed number of Norovirus is not significantly correlated with raw water pH, rainfall and season type (i.e., winter season). The rest of the input variables (turbidity, conductivity and temperature) however show considerable negative correlations with the virus count.

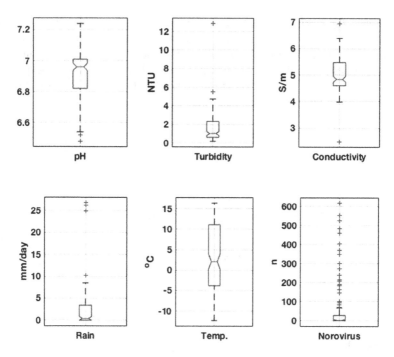

Fig. 3. Boxplot of the raw data

Table 1. Correlation and dependence between Norovirus and water quality variables

Parameter	Correlation	P-value
pH	0.0383	0.6346
Turbidity	**−0.2084**	**0.0090**
Conductivity	**0.3908**	**0.0000**
Rain	−0.0426	0.5975
Temperature	**−0.3048**	**0.0001**
Winter season	0.1057	0.1889

6.1 Prediction of NoV Using GPML

The main goal of this study is to assess the applicability of two modeling approaches for real time prediction of the count of Norovirus in raw water using water quality variables. First, the response of the GPML model is plotted against the normalized measured Norovirus concentration in the raw water, as it is shown in Fig. 4 (upper) while prediction error is shown in Fig. 4 (bottom). The model adequately captures the variations in the counts of Norovirus in the raw water. The model predictions are however not very close to actual observations in most cases. For instance, between January and April 2011, where most of the high counts of the virus were observed in the raw water, less than half of the normalized counts are mostly predicted by the model. The model accordingly underestimates the levels of the virus in the raw water. For a model to be applicable at the industrial scale for real time estimation of pathogenic organisms in raw water, its predictions should be close to actual levels. The mean absolute and mean squared prediction errors are 0.3614 and 0.4179, respectively. The ranges of the model prediction errors, as shown in the lower section of Fig. 4, are the close to the ranges of the observations, suggesting low overall performance.

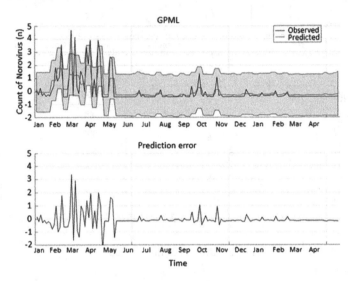

Fig. 4. Response of GPML model in red plotted against the normalized measured Norovirus concentrations in raw water source while the marginal likelihood is shown in gray color (upper) and prediction error (bottom). (Color figure online)

6.2   Prediction of NoV Using ANFIS

Various settings of ANFIS model are used for predicting NoV count as follows:

6.2.1   6 Inputs 2 MFs ANFIS Model (ANFIS6, 2)

$\text{ANFIS}_{6,2}$ has six inputs and 2 MFs namely *low* and *high* to fuzzy each crisp input; pH, turbidity, conductivity, rainfall, temperature, and time accounting for seasonality and one output, the concentration of Norovirus.

In this model, each input is represented by two fuzzy sets, and the output is represented as a first-order polynomial of the inputs. The ANFIS extracts $r = 2^6 = 64$ rules mapping the water quality variables to Norovirus count from the input/output observations. The proposed $\text{ANFIS}_{6,2}$ is shown in Fig. 5. The bell shape membership functions (MFs) were adjusted using gradient vector computation. The measured values of the water quality variables are adjusted to establish their relationships with the count of Norovirus using neural network. The errors associated with the training are monitored using the mean square error (MSE). The optimal number of epochs for the model with the least MSE was found by carrying out numerous experiments. This involved setting different number of epochs during the model training, enabling us to determine the optimal epoch with the least MSE, and therefore avoiding model overfitting. The initial and final MFs for the input variables are shown in Figs. 6 and 7, respectively, where the best result was obtained after 250 epochs with a MSE = 0.3567 for the normalized output. A sample of the generated rules is shown in Table 2.

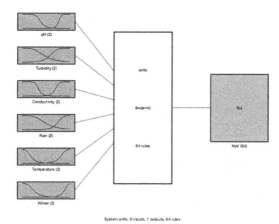

Fig. 5. ANFIS6, 2 for predicting Norovirus count, y, in terms of pH, turbidity, etc. as the model inputs.

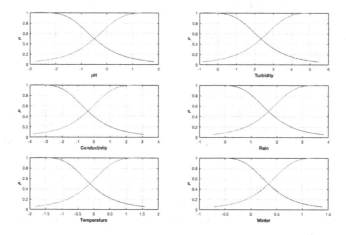

Fig. 6. Initial MFs of the proposed ANFIS6, 2 model (2 MFs are used for fuzzifying each input).

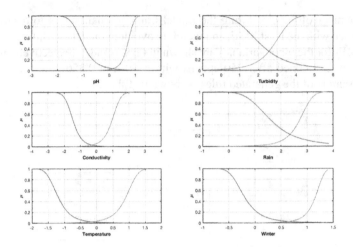

Fig. 7. Final MFs of the proposed ANFIS6, 2 model after 250 epochs of training.

Table 2. ANFIS$_{6,2}$ generalized fuzzy rules after 250 epochs of training

Rule	Rule's description
1	**IF** *pH* is low **AND** *turbidity* is low **AND** *conductivity* is low **AND** *rain* is low **AND** temperature is low **AND** winter is low **THEN** $y = 0.43 + 0.16pH - 0.23turbidity + 0.06$conductivity $- 0.39$rain $+ 0.40$temperature $- 0.56$winter
2	**IF** *pH* is low **AND** *turbidity* is low **AND** *conductivity* is low **AND** *rain* is low **AND** temperature is low **AND** winter is high **THEN** $y = 0.54pH - 0.25turbidity + 0.36$conductivity $+ 0.60$rain $+ 0.09$temperature $+ 0.07$ winter
...	...
64	**IF** *pH* is high **AND** *turbidity* is high **AND** *conductivity* is high **AND** *rain* is high **AND** temperature is high **AND** winter is high **THEN** $y = 0.92 + 0.91pH - 0.06turbidity + 0.51$conductivity-$0.13$rain $+ 0.96$temperature $- 0.69$winter

To assess the predictability of the proposed model, the response of the ANFIS model with all the six inputs (ANFIS$_{6,2}$) is plotted against the normalized measured Norovirus concentration in the raw water, as it is shown in Fig. 8 (upper) while prediction error is shown in Fig. 8 (bottom).

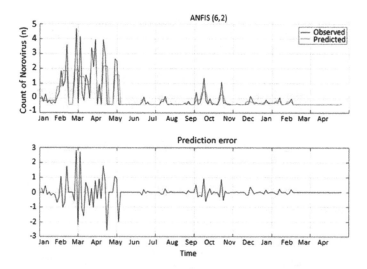

Fig. 8. Response of ANFIS6, 2 model and its mean squared prediction error

The model was capable predicting periods in which counts of Norovirus were observed in the raw water as well as periods of no counts. More importantly, the model efficiently predicted periods of intense variations in the counts of the virus in raw water (the first three months of the study period). This is a necessary information for the optimization of water treatment processes in order to prevent potential waterborne illnesses. Comparing the range of the normalized counts of Norovirus in the models predictions to the associated errors (Fig. 8), it is evident that the model performs better than the GPML model. Further, the ANFIS model predicts zero counts of the virus with higher precision.

In addition, to examine how interactions among the water quality variables affect the level of Norovirus in the raw water, surface view of the input-output mapping were generated as shown in Fig. 9. It is evident from these plots that the influence of each water quality parameter on the virus differs when it interacts with a different variable. For instance, while high pH at high temperature are associated with increased counts of the virus in the raw water, elevated water turbidity occurring at high pH results in lower counts. Similarly, high turbidity result in increasing the level of the virus when conductivity is also high. However, for the same turbidity level, increasing temperature results in lower counts of the virus in the raw water. Finally, interactions among certain pairs of variables (such as between turbidity and pH, turbidity and conductivity) appear to have higher impact on the number of Norovirus than interactions among other pairs (e.g., conductivity and pH).

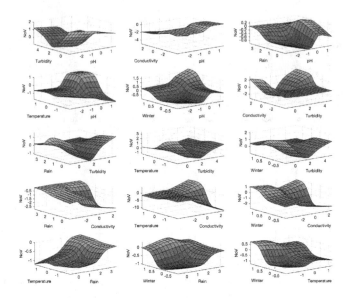

Fig. 9. Surface view of the generated ANFIS$_{6,2}$ fuzzy relations showing the effect of each input variable to the count of the predicted NoV.

6.2.2 3 Inputs 2 MFs ANFIS Model (ANFIS3, 2, COR)

We further conducted an experiment to determine if significant improvement in the performance of the ANFIS model can be achieved with fewer inputs. A reduced set of the input features; turbidity, conductivity, and temperature were accordingly used to train this model as they are more relevant to the concentration of Norovirus, as it was shown in the correlation coefficients in Table 1. Two MFs namely *low* and *high* to fuzzy each crisp input is developed. After 250 epochs, the model was able to extract $2^3 = 8$ rules with prediction error MSE = 0.3735. No significant reduction in the prediction

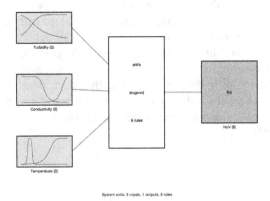

Fig. 10. ANFIS3, 2, COR for predicting Norovirus count, Nov, in terms of turbidity, conductivity and temperature.

error was achieved. Moreover, compared to the full model, the reduced model yielded Norovirus counts at periods where zero counts were actually observed in the raw water. The network structure of the developed $ANFIS_{3,2,COR}$ model using two MFs and 8 rules is shown in Figs. 10 and 11.

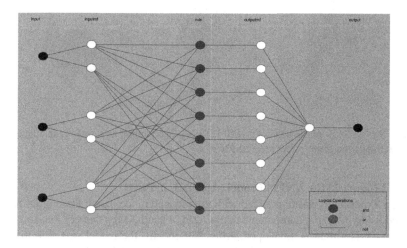

Fig. 11. ANFIS model structure for three inputs and 2 MFs each.

The model predictions versus target values are shown in Fig. 12 (upper) while prediction error is shown in Fig. 12 (bottom). Surface view of the input-output mapping of this model is shown in Fig. 13. Here, the effects of the interactions among the water quality parameters found to be more significantly correlated with the count of Norovirus (turbidity, conductivity, temperature) are more distinct. For instance, a sharp drop in the

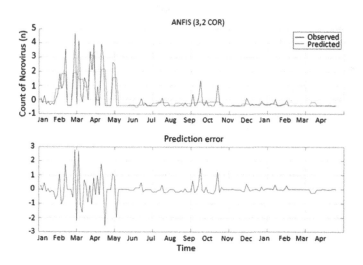

Fig. 12. Response of $ANFIS_{3,2,COR}$ model and the mean squared error.

count of the virus resulting from a combined increases in water turbidity (>2 NTU) and conductivity (>1 μS/cm) can be seen in Fig. 12.

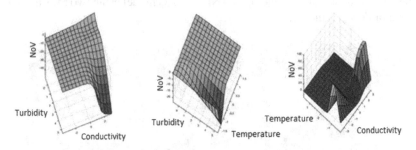

Fig. 13. Surface view of the generated ANFIS$_{3, 2, COR}$ fuzzy relations showing the effect of each input variable to the count of the predicted NoV.

6.2.3   3 Inputs 2MFs ANFIS Model (ANFIS3, 2, PCA)

A further experiment was conducted with the reduced model using three input features obtained using principal component analysis (PCA) and 2 MFs namely *low* and *high* to fuzzy each crisp input is developed. After 250 epochs, the model was able to extract $2^3 = 8$ rules with prediction error MSE = 0.3623. The model predictions versus target values are shown in Fig. 13 (upper) while prediction error is shown in Fig. 14 (bottom). No distinct improvement was observed in the model with PCA input selection over the reduced model with inputs drawn from the correlation analysis. The full model with six inputs marginally performs better than the two reduced models. In addition, the reduced ANFIS model from the PCA-selected inputs poorly explained the relationships among the water quality variables and their associated effects on variations in the count of the

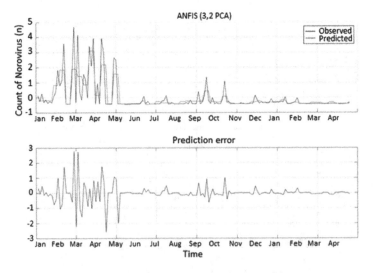

Fig. 14. Response of ANFIS$_{3, 2, PCA}$ model and the mean squared error.

virus in raw water. This is evident from the surface view of the input-output mapping of this model shown in Fig. 15.

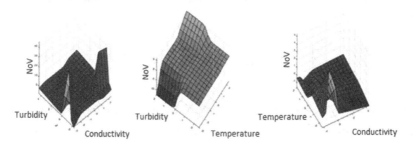

Fig. 15. Surface view of the generated ANFIS$_{3, 2, PCA}$ fuzzy relations showing the effect of each input variable to the count of the predicted NoV.

6.2.4   6 Inputs 3 MFs ANFIS Model (ANFIS6, 3)

Finally, after all the previous ANFIS experiments which involved two membership functions (*'low'* and *'high'*) for each input variable, another ANFIS model was experimented using 6 inputs and 3 MFs, namely, *'low'*, *'medium'*, and *'high'* for fuzzifying each input variable. After 250 epochs, the model generated $3^6 = 729$ rules and a prediction error MSE = 0.3567. This shows that based on the dataset used in this study, it is difficult to achieve higher performance of the ANFIS model, irrespective of the structure. Also, it was observed that increasing the number of membership functions for each input does not appear to have any significant effect on the model performance. Accordingly, after the various ANFIS experiments, the model with six water quality variables as inputs, and with two membership functions (Generalized bell-shaped) produced the best fit at 250 epochs.

The results demonstrate that the ANFIS model can be reliably applied at full scale water treatment facilities for real time prediction of potential concentrations of Norovirus in raw water using measurements of water quality parameters. Detection of pathogens in water has a lot of challenges, including cost, time, operator skills and uncertainties. Due to these challenges, various modeling techniques have been widely applied with the aim of predicting concentrations of the organisms in raw water in real time. However, model inconsistencies, under and over-estimations of actual levels as well as adaptability of the models make them unreliable for application at the industrial scale. The ANFIS model as applied in this study considerably overcomes these challenges, and can be fully applied as part of regular water quality monitoring exercise at the Nødre Romrike Vannverk water treatment plant in Oslo, Norway.

6.3   Model Comparison

In this study, both the GPML and ANFIS models of various structures were trained on the entire dataset and tested for predicting the concentration of Norovirous in raw water source using climate and water quality parameters that are measured at water treatment plants in real time. The performances of both models were compared by the mean absolute error (MAE), the mean squared error (MSE), and R^2 criteria are shown in Tables 3 and 4. The MAE indicates how close the predictions are to the measured values which is given by:

$$MAE = \frac{1}{n} \sum_{i=1}^{n} |f_i - y_i| = \frac{1}{n} \sum_{i=1}^{n} |e_i| \tag{8}$$

Table 3. Performance comparison of GPML and ANFIS models for the test data set.

Model	Inputs	MFs	Rules	MAE	MSE
GPML	-	-	-	0.3614	0.4179
ANFIS$_{6,3}$	6	3	729	0.2704	0.3567
ANFIS$_{6,2}$	6	2	64	0.2704	0.3567
ANFIS$_{3,2,PCA}$	6	2	8	0.2917	0.3623
ANFIS$_{3,2,COR}$	3	2	8	0.299	0.3735

Table 4. R^2 criteria comparison of GPML and ANFIS models.

Modelling technique	R^2
GPML	0.7802
ANFIS$_{6,3}$	0.8006
ANFIS$_{6,2}$	0.8006
ANFIS$_{3,2,PCA}$	0.7971
ANFIS$_{3,2,COR}$	0.79

As seen in Eq. (8), the mean absolute error can be defined as the average of absolute errors; the absolute error given by $|e_i| = |f_i - y_i|$, where f_i is the prediction and y_i the true value. It should be noted that in MAE, all the individual errors have equal weight in the average, making it a linear score. In order to have a reliable statistical comparison between the mathematical models, both the MAE and MSE can be used together to ascertain the variation in errors in a given set of predictions. Calculation of MSE involves squaring the difference between the predicted and corrsponding observed values, and averaging it over the sample size. This can be written as:

$$MSE = \frac{1}{n} \sum_{i=1}^{n} e_i^2 \tag{9}$$

MSE has a quadratic error rule, where the errors are squared before being averaged. As a result, a relatively high weight is given to large errors. This could be useful when large errors are undesirable in a statistical model.

From Table 3 it can be deduced that for the Gaussian model the MSE is slighly higher as Compared to ANFIS. Another measure of goodness-of-fit of the model is the R^2

criteria. R^2 is a proportion of variance '*explained*' by the model. Higher values are indicative that the predictive model fits the data in a better way. By definition, R^2 is the proportional measure of variance of one variable that can be predicted from the other variable. Thus, ideally, the values of R^2 to approach one is always desirable.

However, a high R^2 tells you that the curve came very close to the points but in reality it does not always indicate the model quality. From Table 4, both Gaussian and ANFIS models have similar R^2 values which indicate that in both modeling techniques, the prediction capability is similar. However, using the R^2 criteria in conjunction with the MAE and MSE, it can be fairly deduced that the Gaussian and ANFIS models can be accurately used for the prediction of Norovirous concentrations in raw water source.

7   Conclusions

ANFIS and GPML models for prediction of the counts of Norovirus in raw water have been developed and the predictive abilities of the two models compared. Both modeling approaches have demonstrated adequate performances in their predictions during the model testing stage. In terms of the mean square error and mean absolute error values of the model predictions, the ANFIS models showed considerable accuracy relative to the GPML model. However, the computed R^2 values of the models indicate that no distinct disparity exists between the model performances. However, certain features of the ANFIS modelling approach make it more efficient for application. For instance, the advantages of ANFIS compared to other black box models include: (1) it combines in a transparent manner the linguistic representations of fuzzy logic and the learning capabilities of artificial neural networks, (2) it provides an automated approach for rule generation and parameter optimization procedure that simplifies the complex process of model development, and finally (3) it creates a transparent solution that is expected to offer useful insights into the physical processes involved in the modeling process and therefore help its end user the understand why certain values are obtained.

PCA in this paper is used to provide a reduced set of input features where the most influential three variables to NoV count are chosen. Although the ANFIS model trained using this reduced set of variables is able to provide better predictions, it lacks the ability to provide a meaning to the input-output mapping compared to the ANFIS model trained using the reduced set obtained from correlation coefficients table. It is also noted that using more than 2 MFs does not have any significant effect on the predictability of the produced ANFIS model. For future work, new models and new data sets will be used for providing more accurate predictions of NoV count in terms of climate and water quality parameters.

Acknowledgements. The authors wish to thank the managers of the Nødre Romrike Water Treatment Plant in Oslo for the provision of required data. Thanks to Ricardo Rosado and Mette Myrmel for providing the Norovirus data. This work is part of the project KLIMAFORSK funded by the Research Council of Norway (Project No: 244147/E10). The authors would like to express their sincere thanks to the editor and anonymous reviewers for their suggestions and comments to improve the quality of the paper.

References

1. Ahmed, S.M., Hall, A.J., Robinson, A.E., Verhoef, L., Premkumar, P., Parashar, U.D.: Global prevalence of norovirus in cases of gastroenteritis: a systematic review and meta-analysis. Lancet Infect. Dis. **14**(8), 725–730 (2014)
2. Brion, G.M., Neelakantan, T.R., Lingireddy, S.: Using neural networks to predict peak Cryptosporidium concentrations. J. Am. Water Works Assoc. (AWWA) **93**(1), 99–105 (2001)
3. Bartsch, S.M., Lopman, B.A., Ozawa, S., Hall, A.J., Lee, B.Y.: Global economic burden of norovirus gastroenteritis. PLoS One **11**(4), e0151219 (2016). https://doi.org/10.1371/journal.pone.0151219
4. Altintas, Z., Gittens, M., Pocock, J., Tothill, I.E.: Biosensors for waterborne viruses: detection and removal. Biochimie **115**(2015), 144–154 (2015)
5. Wigginton, K.R., Kohn, T.: Virus disinfection mechanisms: the role of virus composition, structure, and function. Curr. Opin. Virol. **2**(1), 84–89 (2012)
6. Xagoraraki, I., Yin, Z., Svambayev, Z.: Fate of viruses in water systems. J. Environ. Eng. **140**(7), 04014020 (2014)
7. da Silva, A.K., Le Saux, J.C., Parnaudeau, S., Pommepuy, M., Elimelech, M., Le Guyader, F.S.: Evaluation of removal of noroviruses during wastewater treatment, using real-time reverse transcription-PCR: different behaviors of genogroups I and II. Appl. Environ. Microbiol. **73**(24), 7891–7897 (2007)
8. Laverick, M.A., Wyn-Jones, A.P., Carter, M.J.: Quantitative RT-PCR for the enumeration of noroviruses (Norwalk-like viruses) in water and sewage. Lett. Appl. Microbiol. **39**(2), 127–136 (2004)
9. Westrell, T., Teunis, P., van den Berg, H., Lodder, W., Ketelaars, H., Stenström, T.A., de Roda Husman, A.M.: Short-and long-term variations of norovirus concentrations in the Meuse River during a 2-year study period. Water Res. **40**(14), 2613–2620 (2006)
10. Barrett, M., Fitzhenry, K., O'Flaherty, V., Dore, W., Keaveney, S., Cormican, M., Clifford, E.: Detection, fate and inactivation of pathogenic norovirus employing settlement and UV treatment in wastewater treatment facilities. Sci. Total Environ. **568**, 1026–1036 (2016)
11. Lodder, W.J., de Roda Husman, A.M.: Presence of noroviruses and other enteric viruses in sewage and surface waters in The Netherlands. Appl. Environ. Microbiol. **71**(3), 1453–1461 (2005)
12. Ueki, Y., Sano, D., Watanabe, T., Akiyama, K., Omura, T.: Norovirus pathway in water environment estimated by genetic analysis of strains from patients of gastroenteritis, sewage, treated wastewater, river water and oysters. Water Res. **39**(18), 4271–4280 (2005)
13. Chen, H., Hu, Y.: Molecular diagnostic methods for detection and characterization of human noroviruses. Open Microbiol. J. **10**(1), 78–89 (2016)
14. Lermontov, A., Yokoyama, L., Lermontov, M., Machado, M.A.S.: River quality analysis using fuzzy water quality index: Riberia do Iguape river watershed, Brazil. Ecol. Indic. **9**(2009), 1188–1197 (2009)
15. Andreas, T., Olof, B., Bertil, F.: Precipitation effects on microbial pollution in a river: lag structures and seasonal effect modification. PLoS One **9**(5), e98546 (2014)
16. Bruggink, L.D., Marshall, J.A.: Norovirus epidemics are linked to two distinct sets of controlling factors. Int. J. Infect. Dis. **13**(2009), e125–e126 (2009)
17. Sokolova, E., Pettersson, T.J.R., Bergstedt, O., Hermansson, M.: Hydrodynamic modelling of the microbial water quality in a drinking water source as input for risk reduction management. J. Hydrol. **497**(2013), 15–23 (2013)
18. Icaga, Y.: Fuzzy evaluation of water quality classification. Ecol. Ind. **7**(2007), 710–718 (2007)

19. Petterson, S.R., Stenström, T.A., Ottoson, J.: A theoretical approach to using faecal indicator data to model norovirus concentration in surface water for QMRA: Glomma River, Norway. Water Res. **91**, 31–37 (2016)
20. Marshall, J.A., Bruggink, L.D.: The dynamics of norovirus outbreak epidemics: recent insights. Int. J. Environ. Res. Public Health **8**(4), 1141–1149 (2011). https://doi.org/10.3390/ijerph8041141
21. Mohammed, H., Hameed, I.A., Seidu, R.: Adaptive neuro-fuzzy inference system for predicting norovirus in drinking water supply. In: International Conference on Informatics, Health & Technology (ICIHT), pp. 1–6. IEEE (2017)
22. Bisht, D.C.S., Jangid, A.: Discharge modelling using adaptive neuro-fuzzy inference system. Int. J. Adv. Sci. Technol. **31**(2011), 99–114 (2011)
23. Chowdhury, S., Champagne, P., McLellan, P.J.: Models for predicting disinfection by product (DBP) formation in drinking waters: a chronological review. Sci. Total Environ. **407**(14), 4189–4206 (2009)
24. Heddam, S., Bermad, A., Dechemi, N.: ANFIS-based modelling for coagulant dosage in drinking water treatment plant: a case study. Environ. Monit. Assess. **184**(4), 1953–1971 (2012)
25. Sahu, M., Mahapatra, S.S., Sahu, H.B., Patel, R.K.: Prediction of water quality index using neuro fuzzy inference system. Water Qual. Exposure Health **3**(3–4), 175–191 (2011)
26. Jang, J.S.R.: ANFIS: adaptive network-based fuzzy inference systems. IEEE Trans. Syst. Man Cybern. **23**(1993), 665–685 (1993)
27. Negnevitsky, M.: Artifial Intelligence: A Guide to Intelligent Systems, 3rd edn., pp. 277–285. Pearson (2005)
28. VISK. http://www.norskvann.no/, http://www.nrva.no/, http://visk.nu/. Accessed Oct 2016
29. Grøndahl-Rosado, R.C., Tryland, I., Myrmel, M., Aanes, K.J., Robertson, L.J.: Detection of microbial pathogens and indicators in sewage effluent and river water during the temporary interruption of a wastewater treatment plant. Water Qual. Exposure Health **4**(3), 155–159 (2014)
30. Rasmussen, C.E., Williams, C.K.I.: Gaussian Processes for Machine Learning. The MIT Press, Cambridge (2006)
31. Nickischm, H., Rasmussen, C.E.: Approximations for binary Gaussian process classification. J. Mach. Learn. Res. **9**, 2035–2078 (2008)
32. Rasmussen, C.E., Williams, C.K.I.: Gaussian Process for Machine Learning. The MIT Press (2006). ISBN 026218253X, Matlab code version 4.0 http://gaussianprocess.org/gpml/code/matlab/doc/index.html. Accessed 15 Apr 2017

Cloud Computing Adoption in Healthcare Organisations: A Qualitative Study in Saudi Arabia

Fawaz Alharbi[1,2(✉)], Anthony Atkins[1], and Clare Stanier[1]

[1] School of Computing and Digital Technologies, Staffordshire University, Stoke-on-Trent, UK
fawaz.alharbi@research.staffs.ac.uk,
{a.s.atkins,C.Stanier}@staffs.ac.uk
[2] Huraymila College of Science and Humanities, Shaqra University, Shaqra, Saudi Arabia

Abstract. This paper provides a comprehensive review of Cloud Computing by discussing the benefits and challenges of implementing such solution and discusses various Cloud Computing adoption models. The paper describes Cloud Computing in healthcare domains. It provides also information about Cloud Computing in Saudi Arabia and how it could be applied for healthcare domain. The paper presents a qualitative study which provides an in-depth understanding of the Cloud Computing adoption decision-making process in healthcare organisations in Saudi Arabia. The paper discusses the factors which will affect Cloud Computing decision making process in Saudi Arabia. The findings of the study showed that the factors affecting Cloud Computing adoption can be divided into five main categories, Technological, Business, Environmental, Organisational and Human. This paper also identifies some of the key drivers and challenges of Cloud Computing adoption in Saudi healthcare organisations. This study will help both Saudi healthcare organisations and Cloud Computing vendors in understanding healthcare organisations' attitude towards the adoption of Cloud Computing.

Keywords: Cloud computing · E-health · Qualitative study · Saudi Arabia · Strategic framework

1 Introduction

Healthcare organisations around the world try to employ innovative and effective solutions to provide better healthcare services. One such initiative is the use of information and communication technology in health organisations to deliver healthcare more efficiently and effectively. The use of IT in the healthcare sector in general is referred to as e-health [1]; e-health can provide or facilitate many benefits such as enhanced information sharing, improved healthcare quality and improved healthcare safety. However, current e-health solutions and projects face some challenges such as the high cost of implementing IT systems in healthcare services [2], the shortage of health informatics specialists and IT professionals [1], the presence of heterogeneous devices [1] and the large amount of data in healthcare applications [3]. This creates a foundation for the use of innovative technologies and models that may move healthcare IT systems forward,

© Springer-Verlag GmbH Germany 2017
A. Hameurlain et al. (Eds.): TLDKS XXXV, LNCS 10680, pp. 96–131, 2017.
https://doi.org/10.1007/978-3-662-56121-8_5

such as the adoption of Cloud Computing. Although researchers identify Cloud Computing as an IT global phenomenon, they also highlight that the factors affecting Cloud Computing adoption play different roles across different economic environments. Legal factors such as data protection laws are different between countries even where the countries may be from the same region [4]. Government support for Cloud Computing also varies between countries. While some technologically advanced countries have launched Cloud Computing initiatives such as G-Cloud in UK and the Kasumigaseki Cloud in Japan, other countries such as Saudi Arabia have not yet undertaken a national Cloud Computing initiative. Cloud Computing research in the literature focuses mainly on technologically developed countries and fewer empirical studies have been conducted in developing countries [5]. Cultural and organisational characteristics may affect how different countries adopt Information Technology projects. Examining the impact of cultural and organisational factors across different industries and countries represents a contribution to the body of knowledge about Cloud Computing adoption.

Successful Cloud Computing adoption in the health sector requires strategic planning to take full advantage of this emerging model. Understanding the different strategic aspects and factors of Cloud Computing is important and could encourage organisations to adopt this model of computing [6].

This paper discusses a qualitative approach used with senior decision-makers in hospitals in Saudi Arabia to identify the factors affecting Cloud Computing in Saudi Healthcare organisations. The rest of the paper is organised as follows: Sect. 2 reviews Cloud Computing and its advantages and issues, Sect. 3 discusses Cloud Adoption Decision Making Models, Sect. 4 shows the application of Cloud Computing in healthcare domain and the related work, Sect. 5 discusses Cloud Computing in the Saudi Arabia context and particularly in the healthcare domain, Sect. 6 discusses the research approach, Sect. 7 discusses the conduct and findings from the study and Sect. 8 discussed the implications of the findings and Sect. 9 gives the conclusions and identifies future work.

2   Cloud Computing

The following sections provide an overview of Cloud Computing and describe the advantages and challenges of implementing Cloud Computing solutions.

2.1   Overview of Cloud Computing

Cloud Computing could be seen as paradigm that has evolved from previous computing paradigms. [7] suggested that there are six distinct phases for computing development as shown in Fig. 1.

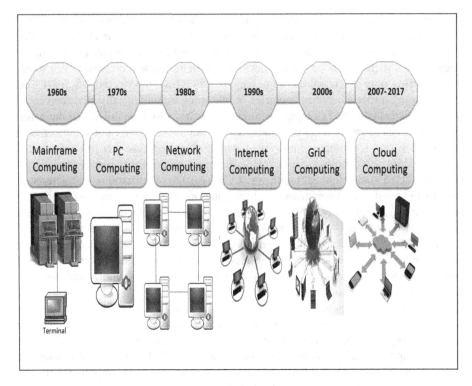

Fig. 1. Computing paradigm shift

The National Institute of Standards and Technology (NIST) has defined Cloud Computing as "a model for enabling ubiquitous, convenient, on-demand network access to a shared pool of configurable computing resources (e.g., networks, servers, storage, applications, and services) that can be rapidly provisioned and released with minimal management effort or service provider interaction" [8]. A Cloud service model represents the services and the capabilities which will be offered via the Cloud. The fundamental service models based on NIST definition are Software as a Service (SaaS), Platform as a Service (PaaS), Infrastructure as a Service (IaaS) [8]. A Cloud Deployment Model refers to the type of Cloud environment with variations in management, ownership, access and physical locations [8]. Based on the NIST definition [8], there are four deployment types which are Public Cloud, Private Cloud, Hybrid Cloud and community Cloud.

2.2 Advantages of Cloud Computing

In the following sections, we discuss the advantages of Cloud Computing under three headings, Technical, Economic and Organisational factors.

Technical Benefits

Technical benefits are identified as important drivers for organisations to adopt Cloud Computing (Carroll et al. 2011). Scalability was identified as the first driver for Cloud Computing adoption in North Bridge Future of Cloud Computing Survey in 2011, 2015 and 2016 [9] and as the second most important factor in other reports [10, 11]. Cloud Computing improves the IT agility of organisations since organisations can reduce the time between identifying a need for a new IT resource and delivering it [12]. A Harvard Business review report identified this capability of the Cloud as one of the main drivers for Cloud Computing adoption [13]. Cloud Computing can offer organisations access to cutting-edge IT resources and these resources are not required to be available physically in the organisations [14]. 70% of the organisations surveyed were satisfied with the ability of Cloud Computing to grant their organisations access to the best and newest technologies [15]. Cloud Computing delivers increased flexibility in the use of IT resources and improves access by increasing system availability (up time) [16]. Cloud Computing could also allow IT department to gain access to specialized technical skills and cloud vendor services, leading to 'value propositions' for organisations [17]. Cloud service providers can improve the power efficiency of their data centres by 40% compared to traditional data centres [18].

Economic Benefits

The financial and economic benefits of Cloud Computing can be quantified in terms of money either generated or saved. Economic considerations have been found to have significant influence on an organisation's decision to adopt Cloud Computing. Factors related to business are considered as the main drivers for Cloud Computing adoption in many surveys [19] with cost savings identified as one of the main reasons [4, 20]. The North Bridge 2011 Future of Cloud Computing Survey showed that cost was the second driver for organisations to move to Cloud Computing and cost is still ranked the third primary driver in the 2016 survey [9]. A study showed that 37% of organisations had the potential to achieve cost saving when implementing Cloud Computing solutions [21]. Total Cost of Ownership (TCO) for IT services in organisations is estimated to go down between 10% to 30% by using Cloud Computing [22].

Organisational Benefits

Cloud Computing offers organisations additional benefits besides the economic and technical benefits. Many studies showed that Cloud Computing can inspire innovation culture in organisations in many ways [20]. 50% of the respondents to the Harvard Business Review survey clarified that Cloud Computing increases their organisation's ability to innovate [23]. Cloud Computing also frees an organisation's IT staff to spend more time on strategic initiatives and innovation. 60% of surveyed USA and UK organisations reported that Cloud Computing allowed their IT staff to focus more on strategy and innovation activities instead of operational and maintenance activities [24]. The ubiquitous nature of the cloud offers more flexibility for employees to work independently from any device and location [25]. Cloud Computing helps organisations to provide new services that were not possible before due to previously higher costs for IT solutions [4]. This was found to be one of the main benefits of Cloud Computing adoption in 28% of surveyed UK organisations [11]. The studies also indicated that the adoption of Cloud Computing will improve collaboration between an organisation's employees due to use of features such as mobile access and version control [26]. The Harvard Business Review report indicated that increased collaboration was the top benefit of implementing Cloud solutions [13].

From the literature and industrial reports, technical and economic benefits are identified as the main drivers of Cloud Computing adoption. However, organisational benefits such as delivering new services and applications, ability to react quickly to changing market conditions and enabling innovation have been recognised in many studies and the strategic values of Cloud Computing increases with the maturity of Cloud Computing. A 2016 IDG study identified organisational benefits as among the most highly cited drivers of Cloud Computing implementation [10].

2.3   Challenges of Cloud Computing

In the following sections we discusses the challenges of Cloud Computing under three headings, Security Concerns, Technical Concerns and Non-Technical Concerns [27].

Security Concerns

Several studies identify security concerns as one of the main challenges of Cloud Computing adoption [19, 28]. The North Bridge Future of Cloud Computing Survey in 2012 showed that 55% of interviewed experts and users consider security challenges as an inhibitor for Cloud Computing adoption [29]. However, this percentage decreased to 38.6% in the 2016 survey [9] probably as adoption increases. A study conducted by Phaphoom et al. (2015) found that security challenges were perceived as a higher concern for non-adopters of Cloud Computing [28]. A study conducted by SAP and Oxford Economics showed that 40% of IT leaders have some concerns about unauthorized access to sensitive data [30]. Although it has been argued that data integrity, protecting the data from unauthorised alterations [31], can be maintained effectively in Cloud Computing environment via Secure Digital Signatures tools [32], some Cloud technologies such as virtualisation can lead to additional risks [33]. Data privacy related to security has also been identified as a challenge for Cloud Computing adoption [28]. According to a KPMG Cloud report, 53% of the participants clarified that data loss and privacy risks are the challenges that affecting their organisations implementation of Cloud solutions [34]. Cloud adopters are able to address some of these concerns via better governance, security practices and robust Service Level Agreement (SAL) and proven adoption [32, 35]. It has been argued that leading Cloud providers are able to provide more advanced data security measures than some organisations [36]. A Right-Scale report reported that the top challenge for adopting Cloud Computing is not security and security challenges were cited only by 32% percent of the respondents [37]. 99% of 250 senior IT and business decision-makers have not experienced any breach of security when implementing Cloud Computing in UK organisations [11].

Technical Concerns
The spread of Cloud Computing solutions has raised some technical concerns [28]. The technical concerns of Cloud Computing implementation can be divided into sub categories such as Integration with existing IT infrastructure, Reliability and performance of Cloud solutions and Vendor lock-in. For some organisations, adopting Cloud Computing will require additional effort in terms of configuration management to ensure compatibility and integrity [14]. A survey conducted by KPMG showed that Cloud Computing integration with existing IT architecture is found to be challenging by 46% of the survey respondents [34]. Performance was mentioned by only 9% of the participants in a Harvard Business Review report as a barrier to Cloud Computing adoption [13]. However, performance could be affected by other factors such as: internal IT infrastructure and bandwidth [38]. Vendor lock-in was considered by 25% UK organisation as an inhibitor of Cloud adoption [11]. However, there are efforts to provide standardisation and initiatives to create an Inter-cloud environment that support both interoperability and portability [39].

Non-technical Concerns
There are also non-technical issues which may affect Cloud Computing adoption. Examples of non-technical issues include Legal Concerns, Hidden costs, Change resistance and Lack of resources and expertise. Legal and compliance requirements concerns were identified as challenges for organisations planning to adopt Cloud Computing in many studies. For example, legal and regulatory compliance was cited by 46% of the respondents as challenge for adopting Cloud solutions [34]. Organisations planning to adopt Cloud Computing Services could face issues with the hidden costs of adoption which may include application migration costs, human resource costs and integration costs [40]. For example, public cloud providers apply bandwidth charges for outbound data transfers (i.e. data going out of provider's data centres) and the charges could be higher for large data transactions [41]. The review of Cloud Computing concerns showed that as organisations become familiar with Cloud Computing solutions and gain more experience, concerns decline and move from security aspects to other managerial [37].

3   Current Cloud Adoption Decision Making Models

Adopting Cloud Computing is a decision-making problem that raises several critical issues for management so it requires strategic tools and frameworks that support multiple domains. There are various Cloud Computing adoption frameworks and models [42–44] and Cloud Computing related frameworks can be categorised as follows:

- **Cloud Computing Risks and Benefits Assessment Frameworks**

Assessing the benefits and risks of implementing Cloud Computing solutions is an important topic and has been discussed in many studies. Khajeh-Hosseini [42] introduced the Cloud Adoption Toolkit which acted as a benefits and risks assessment tool to help decision makers in identifying the advantages and concerns of adopting public clouds. This tool combined Cost Modeling and Technology suitability analysis techniques to provide decision makers of organisations with initial assessment of benefits and risks of adopting public cloud solutions. However, this tool focused on cost and risk analysis and the technical side of adopting public Cloud Computing without discussing other organisational factors and it is also limited to IaaS model [43, 44]. Azeemi et al. [45] applied the IS Success Model to measure the success of Cloud Computing migration without providing specific measurements or evaluations. Chang et al. [46] developed Cloud Computing Business Framework to resolve issues around Cloud adoption challenges.

- **Selecting Service Providers and Service Orientation**

The selection of service providers is also supported by Cloud Computing models and frameworks. Garg et al. [47] developed the SMICloud framework which is based on a Service Measurement Index (SMI). This framework allows the Cloud customers to compare different Cloud providers based on specific requirements. Eleven quantifiable indicators such as response time, accuracy and cost were defined. This framework considered the technical aspect only since it deals with effective measurement of Quality of Service (QoS). However, this framework ignores qualitative indicators such as organisational issues. Ferrer et al. [48] presented a holistic approach to Cloud service and addressed five concerns that will affect the adoption of Cloud computing. The researchers focused on two stakeholders that are service providers and infrastructure providers. Although this study provided a holistic approach, it is targeted towards vendor services. Whaiduzzaman et al. [49] studied the selection of Cloud service providers and discussed eleven Multiple Criteria Decision Analysis (MCDA) techniques in detail and presented examples of the use of those techniques in Cloud computing. The study is well documented but it focuses on the implementation level and mainly on the technical aspect. Other important aspects such as regulations and higher strategic level point view have not been presented.

- **Cloud Computing Decision Support Systems**

Cloud Computing Decision Support Systems (DSSs) aim to support Cloud migration decision making by automating information collection and decision making processes [44]. However, most of the frameworks in this category focus on supporting migration processes such as selection of providers and services and require detailed information to support the selection process [44]. Manzel and Ranjan [50] proposed CloudGenius to automate the decision-making process and support the selection of cloud providers and application migration. Alkhalil [44] provided a decision making model which focused on technical aspects of Cloud migration. Alhammadi [43] developed a Knowledge Management Based Cloud Computing Adoption Decision Making Framework (KCADF) to support processes such as Cloud Adoption Decision Model, Cloud Deployment Selection and Cloud Service Model Selection. However, this framework is still in prototype design and did not include organisational factors such as top management support when evaluating the criteria focused on the Cloud Computing migration process.

- **Cloud Application Migration**

Studying the Cloud migration process is another active research area. Kundra [51] has suggested a decision framework for Cloud migration. This paper presents a strategic perspective for US agencies in terms of consideration and planning for Cloud migration. However, this framework is a high level framework and focused on the technical side of migration decision-making process and is limited to the US government sector. Alonso et al. [52] presented a cloud modernization assessment framework to support the legacy application migration process of Cloud Computing by analysing two perspectives which are technical and business. Although this framework provided a technical and business feasibility analysis and maturity assessment tool, it only considered two contexts of the organisations and focused only on legacy applications. Jamshidi et al. [53] developed the Cloud-RMM migration reference model to support systematic migration to the cloud. However, the framework is categorised as a theoretical framework with an absence of systematic procedures for the implementation [44].

- **Assessment of Organisational Readiness**

In the literature, researchers introduce methods to determine an organisation's Cloud readiness. Loebbecke et al. [54] presented the Magic Matrices Method as a Cloud Readiness assessment tool. The Magic Matrices Method focuses on the operational level of the organisation by investigating selected IT services. The Cloud Computing Assessment Criteria used in this method were (1) Core Business/Competitive Position, (2) Importance/Availability, (3) Standardisation, (4) Degree of Distribution within the organisation, (5) Network Connectivity, (6) Identity Management, and (7) Compliance. The IT services which were assessed by this method were allocated to three categories which are (1) ready for Cloud, (2) not ready or (3) ready in the next few years. Although this method provides an in depth understanding of the technological side of Cloud Computing adoption, it focuses more on the operational level and ignores the strategic level of decision making.

Kauffman et al. in [55] proposed A Metrics Suite for Firm-Level Cloud Computing Adoption Readiness. The Metrics Suite has four main categories which are technology and performance, organization and strategy, economics and valuation and regulation and environment. The metrics suite provided a proposed tool without detailing the implementation. The researchers also did not include factors such as: the attitude towards Cloud Computing adoption, service level agreements and soft financial analysis. The tool requires detailed information which makes it unsuitable for strategic level decision making. Idris et al. in [56] developed an Adoption Assessment tool for Cloud Computing adoption based on Cloud Computing Maturity Model. This tool focused on seven categories which are Business and Strategy, Architecture, Infrastructure, Information, Operations, Projects and Organization. This tool focused more on the operational capabilities of organisations and ignored some important aspects of Cloud Computing adoption such as: human, legal and security aspects.

- **Factors in Cloud Computing adoption**

A number of studies have attempted to identify factors affecting Cloud adoption in different domains and countries such as [12, 35, 57]. One limitation of these studies is the reference to factors without discussing how to implement Cloud Computing decision making based on multiple perspectives. A further limitation is that these studies are usually limited to technologically developed countries [5, 43].

- **Industrial Cloud Computing Adoption Frameworks and models**

Commercial Cloud providers offers tools to support Cloud Computing decision making process such as: Oracle Consulting Cloud Computing Services Framework (OCCCSF), IBM Framework for Cloud adoption (IFCA), and other industrial Cloud Maturity Model [42–44]. One major limitation of such frameworks and models is the difficulties of implementing and adopting such frameworks for non-customers of commercial Cloud providers [42]. The commercial tools are closed proprietary tools which are developed for marketing purposes and require consultancy fees [42].

- **Strategic Cloud Computing frameworks**

This category includes models and frameworks which focus on strategic level decision making of Cloud Computing. Kuo [58] recommended four aspects to be assessed when adopting Health Cloud Computing: management, technology, security, and legal. Kuo also proposed a Healthcare Cloud Computing Strategic Planning (HC2SP) model. This model could act as a SWOT analysis for health organisations to determine how to migrate from traditional health services to cloud-based services and does not focus on the decision making process. Kaisler et al. [59] proposed a decision framework for Cloud Computing to assist small to medium businesses in making decisions about Cloud Computing adoption without detailing the implementation. Qian and Palvia [60] studied Cloud Computing impact on IT strategies and developed The Cloud Impact Model without focusing on decision making process.

Table 1 presents categorisations of current Cloud Computing Adoption Frameworks and Models with brief description about each category and examples of the frameworks.

- **Evaluation of current Cloud Computing decision making frameworks**

Evaluating the existing frameworks for Cloud Computing decision-making shows that these frameworks have number of limitations. The frameworks do not cover multiple perspectives since current models and frameworks focus mainly only on the operational and tactical level (i.e., ad hoc frameworks). Furthermore, while most of the frameworks emphasise the technical side of Cloud Computing, they ignore other elements such as business and organisational. There is a lack of quantitative measures in the reviewed frameworks. The use of quantitative measures is required to make the decision-making process more accurate and objective [61]. Additionally, when some researchers discussed the factors affecting Cloud Computing adoption, they refer to the factors without discussing how to implement them for Cloud Computing decision making based on multiple perspectives. Although some concepts of Cloud Computing will be generic, some of the concepts will be different due to the specific context. Cloud Computing adoption also varies across countries and industries. The evaluation of existing frameworks for Cloud Computing decision-making indicated that there is a need to develop a strategic framework for Cloud computing decision-making processes which emphases a multidisciplinary holistic view of factors affecting Cloud Computing adoption. To develop a strategic understanding of the factors which influence Cloud Computing adoption, it is necessary to examine the views of decision makers.

Table 1. Categorisations of current cloud computing adoption frameworks and models

Category	Description of the category	Frameworks examples
Selecting service providers and service orientation	Frameworks and models related to service provider selection process	[47–49]
Risk and benefit analysis	Assessing the benefits and risks of implementing Cloud Computing solutions	[42, 45, 46]
Cloud computing decision support systems	Systems to aid and support the decision process for migrating to Cloud Computing	[43, 44, 50]
Cloud application migration	Supporting the migration process of Cloud Computing	[51–53]
Assessment of organisational readiness	Methods and tools to determine an organisation's Cloud readiness	[54–56]
Factors in cloud computing adoption	Identifying factors affecting Cloud adoption decision making process in different domain and countries	[12, 35, 57]
Industrial cloud computing adoption frameworks and models	Commercial tools which provided by Cloud vendors to support Cloud Computing decision migration	Oracle Consulting Cloud Computing Services Framework IBM Framework for Cloud adoption
Strategic cloud computing frameworks	Models and frameworks focus on strategic planning of Cloud Computing	[58–60]

4   Cloud Computing in Healthcare

Many healthcare organisations are moving towards new business models and technologies, including Cloud Computing. The Cloud Computing market in healthcare is predicted to reach more than $9.48bn by 2020 [62], which indicates that Cloud Computing will be one of the preferred solutions for healthcare organisations. The following sections present the advantages and concerns of using Cloud Computing solutiohs in healthcare domain and the related work about Cloud Computing adoption in healthcare orgnisations.

4.1   Opportunities Provided by Cloud Computing for E-health

The opportunities and advantages that Cloud Computing offers for e-health [1, 58, 63], are outlined as follows:

Improved Patient Care

Physicians and patients require access to appropriate information to enhance the overall quality of care as current healthcare systems are fragmented so information-sharing tasks are complex and inefficient. A possible reason for this fragmentation is the lack of an information-exchange infrastructure platform [64], which can be provided through Cloud Computing. Allowing physicians to have aggregated information from various sources with historical information about patients' previous healthcare records can enable them to provide more suitable treatments [1]. Large healthcare organisations have already discovered the opportunities that Cloud Computing can provide. For example, the U.S. Department of Health & Human Services (HHS) Office applied a Cloud Shared Services strategy to support their mission and objective goals [65]. Cloud Computing creates the infrastructure for patient information sharing and gives patients the chance to communicate easily with their healthcare providers anytime, anywhere via any device [66]. For example, the Trustworthy Clouds (TClouds) initiative was established by a consortium including IBM, Portuguese and Italian academics, and healthcare organisations to support homecare services and to monitor and assist patients outside the hospital [65]. The MUNICH platform is another project which aims to store and analyse data collected from smart devices in the operating theatres to improve the quality and safety of patient care and to automate the documentation processes [66].

Cost Savings

Healthcare organisations try to deliver high-quality healthcare services with controlled budgets. Cloud Computing can offer economic savings and financial benefits by decreasing the initial and operational costs of e-health projects. Yoo et al. [67] estimated that the return on investment (ROI) of a private Cloud within Seoul National University Bundang Hospital (South Korea) will be 122.6% over a five-year operating period. In China, the 454th Hospital of People's Liberation Army (H454) migrated all its hospital information systems to the Cloud-based VDI platform and the hospital established Cloud-provided hospital information software as a service (HI-SaaS) [68]. Associated smaller healthcare institutions were allowed to share medical software with H545 via SaaS. According to the researchers, 89.9% of the medical clinics that participated saved on investment and maintenance costs [68]. Yao et al. also highlighted that the use of Cloud Computing services can increase resource utilisation and the efficient use of both hardware and software resources. Adopting Cloud Computing solutions allowed the Swedish Red Cross to reduce the costs of IT operations by about 20% and improved real-time communications [69]. Cloud Computing could also reduce the cost of healthcare organisations by allowing them to share IT resources. For example, Liverpool Women's Hospital in the UK and nearby Alder Hey Children's Hospital decided to share their IT resources via private Cloud data centres [70]. Cloud Computing could help at regional or national level by assisting in solving the healthcare information system' fragmentation and isolation problems.

Enhancing Support for Research

Cloud Computing solutions can be used to support and accelerate research and development activities in the healthcare domain [1]. Cloud Computing allows healthcare organisations to have access to powerful computing resources to carry out advanced research activities. For example, Clouds Against Disease is an initiative to support the discovery of new medicines by applying the IaaS model to analyse a trillion possible chemical structures, a task which requires massive computing resources at an acceptable cost [71]. Cloud Computing also makes the process of research and development quicker by empowering the capabilities of healthcare originations [72]. Pfizer, a large pharmaceutical company, combined a commercial Cloud provider's capabilities with the company's High Performance Computing (HPC) facilities, which reduced the computational time to hours instead of weeks [73]. Cloud Computing improves the collaboration of healthcare organisations and different healthcare stakeholders, which can lead to better knowledge sharing [1]. For example, the 100,000 Genomes Project (100,000 GP) is a health project to improve the treatment and diagnosis of patients with rare diseases or cancers where 100,000 genomes are to be collected from the patients and their families [72]. Cloud Computing solutions were implemented to store DNA sequencing, conduct advanced data analysis and allow healthcare professionals and other researchers to benefit from large volumes of data [72].

Allowing a Patient-Centric Healthcare Model

Current healthcare systems are doctor-centred or hospital-centred models where disease and doctors are at the centre [74]. These models focus on the role of physicians and the treatment of illness, which makes this a reactive approach and one that does not respond quickly to the patients' needs [66]. Thus, the movement is towards a model where the patients are the active actors in their healthcare management and this model is called the patient-centric healthcare model [75]. This model empowers individuals to play a larger role in their healthcare and allows for a more proactive and preventive approach [74], and encourages social care and home healthcare services outside of a hospital setting [66]. Cloud computing applications can play an important role in the movement towards a patient-centric model by supporting public health initiatives [66]. For example, self-healthcare management tools that combine Cloud Computing and the

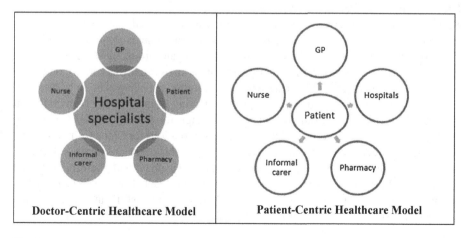

Fig. 2. The transition of the healthcare model from a doctor-centric to a patient-centric model

Internet of Things (IoT) have been implemented to help citizens manage their health status from home [75]. Figure 2 shows the healthcare model's transition from a doctor-centric to a patient-centric model.

Overcome the Issue of Resource Shortages
Cloud Computing offers a possible to route to overcome staff shortages such as the shortage of IT technicians and healthcare professionals in healthcare organisations. Cloud Computing reduces the time spent by IT staff at healthcare organisations on maintenance and operational activities and allows them to work more on strategic tasks and supporting the core business of the organisation [1]. For example, Cloud Computing allowed the Swedish Red Cross to free up to 25% of workers' time, which allowed them to focus more on strategic initiatives and improved collaboration between employees [69]. Cloud Computing can also mitigate the effects of a shortage of healthcare professionals by supporting and improving telemedicine solutions, particular in rural areas [66]. A Cloud Computing solution supported the 12-lead Electrocardiography (ECG) telemedicine implementation in Taiwan which allowed experienced cardiologists to consult with real-time data and enabled urban hospitals to support rural clinics [76]. Cloud Computing helps healthcare organisations use computing resources that were not previously available to them due to the high cost of implementation [1]. The Oshidori-Net2 project applies Cloud Computing solutions to allow six hospitals in Japan to share electronic patient records and PACS solution [77].

4.2   Concerns About Cloud Computing in E-health

Despite the many advantages that Cloud Computing can offer for healthcare organisations, there are still some concerns which may delay its adoption in the healthcare domain. Some of these concerns are related not only to the healthcare organisations but are also relevant to organisations in various other domains. However, certain issues have particular importance in healthcare organisations, which are as follows:

Security and Data Privacy Concerns
Security is a key concern in the implementation of any e-health system and by their nature healthcare organisations have many security requirements. Thus, the implementation of Cloud Computing solutions in healthcare organisations must reflect security and privacy requirements [1]. Virtual infrastructure is an example of a Cloud Computing security risk; patient data could be accessed by unauthorised persons because the hardware resource is used by another person [58]. Protecting patient privacy is an important issue for healthcare organisations and is a challenge for them in Cloud Computing implementation [72]. Therefore, developing secured solutions for Cloud Computing that provides better security and privacy protections for healthcare organisations is an important research area. For example, DACAR is the first e-Health Cloud platform in Europe developed to be a secure platform in the Cloud to support Data Capture and Auto-Identification technology [1]. DACAR has been implemented successfully in a major London hospital [17]. Cloud Computing can potentially provide better security and privacy

practices for some healthcare organisations since they will rely on large Cloud Computing providers that can afford advanced security solutions [58].

Availability and Reliability of Cloud Services
E-health services and applications usually deal with patients, so good availability is required, especially in the event of an emergency. Cloud Computing services may have outages, especially when the services are provided by Cloud providers [65]. However, the overall industry yearly average of uptime for all Cloud providers is 99.999% of uptime, which equals three minutes of unavailability each year. Furthermore, Cloud Computing providers could provide better availability of data than traditional IT operations since they have multiple data centre locations and better backup solutions, which ensures more replication of patients' data.

Regulation Compliance
Governments place great emphasis on protecting patient and medical data and this may be regulated by law. Examples include the US Health Insurance Portability and Accountability Act (HIPAA), the Canadian Personal Information Protection and Electronic Documents Act (PIPEDA) and the UK Data Protection Act of 1998 and Access to Health Records Act 1990 [3, 72]. Healthcare organisations in general are required by law to follow the regulations and it is their responsibility to ensure that their Cloud Computing solutions comply with legislation [78]. Cloud Computing providers recognise the importance of security and privacy requirements and they follow specific measurements that are approved by a third party, such as ISO/IEC standards [79]. The MSSNG project (storing and analysis of DNA of 10,000 families affected by autism) is an example of a Cloud healthcare project which specifies security and privacy standards and measurement [72]. In this project, Google provides the Cloud Computing solutions; Google agreed to implement measures to ensure data privacy and security such as extra layers of encryption, notifying the project management about any security breach, complying with the ISO 27001, SSAE-16, SOC 1, SOC 2, and SOC 3 standards, and storing the project data in Google datacentres in the USA or Europe [72].

4.3   Work Relating to Cloud Computing in E-health

Several studies have discussed Cloud Computing decision-making procedures in healthcare [1]. Kuo [58] recommended four aspects that should be assessed when adopting e-health Cloud Computing: management, technology, security and legal. Kuo also proposed a Healthcare Cloud Computing Strategic Planning (HC2SP) model. As discussed in 3.8, this model could act as a SWOT analysis for migration from traditional health services to cloud-based services but did not focus on the decision-making process. Rijnboutt et al. [63] categorised the challenges facing the use of Cloud Computing in e-health services into six categories (technical, privacy, legal, organisational, economical and medical). However, environmental issues were not considered and the paper did not focus on the decision-making process. Lian, Yen and Wang [80] studied the decision to adopt Cloud Computing in Taiwan hospitals. They integrated the Technology-Organisation-Environment (TOE) framework and Human-Organisation-Technology fit (HOT-fit) model to study the adoption of Cloud Computing in Taiwan. Their study indicated that the five most critical factors in this context are: data security, perceived technical competence, costs, top manager's support and complexity. This study focused on small- and medium-sized hospitals in Taiwan which have a very high degree of e-healthcare maturity [80] and it is not easy to generalise the results of this study to developing countries. The study also did not discuss issues such as technology readiness, change resistance and the availability of external expertise. Gao et al. [81] proposed a framework to evaluate the adoption of Cloud Computing services in Chinese hospitals based on the HOT-fit model. The Gao et al. framework was developed from the literature and from interviews. However, their framework was designed to measure the degree of collaboration between healthcare organisations when adopting Cloud Computing and ignores other aspects of the decision-making process. Harfoushi et al. [82] applied the TOE framework to study the factors affecting Cloud Computing adoption in Jordanian hospitals and found that all three perspectives (i.e. Technology, Organisational and Environmental) affect Cloud Computing adoption. However, their study did not show how these factors impact the decision on the adoption of Cloud Computing and they did not study the sub-factors of each perspective. Osman (2016) examined the factors influencing Adopting Cloud Computing for 9-1-1 Dispatch Centres in the USA and he found that relative advantage, top management support, funding and firm size are the determinants of Cloud Computing adoption. However, the study ignored human and environmental aspects. Lian [83] studied the quality-related factors that influence the successful adoption of Cloud Computing in Taiwanese hospitals based on the information systems success model. This study found that information quality, system quality and trust affected satisfaction with Cloud Computing at the hospitals. However, Lian's study focused only on quality-related factors and did not consider decision-making activities. Some researchers suggest that Cloud Computing in general and in e-health particularly is still in its early prototype stages and needs more research [1, 20, 58, 83, 84]. Although there are many studies and projects about Cloud Computing in the health sector, most of them are focusing on the operational level. Successful Cloud Computing adoption in the health sector requires strategic planning to gain the full advantages of this new model [58].

5   Cloud Computing in E-health in Saudi Arabia

The next sections show an overview of Cloud Computing in Saudi Arabia, summarize Cloud Computing research in Saudi Arabia and discuss Cloud Computing situation for Saudi healthcare organisations.

5.1   Overview of Cloud Computing in Saudi Arabia

Saudi Arabia is one of the largest ICT markets in the Middle East and the total ICT spend in the country is predicted to reach $33.8bn in 2017 [85]. The Cloud Computing market in Saudi Arabia is also expected to grow to $70m in 2017 [86] and is expected to reach $126.9m in 2019 [87]. The Cloud Computing market is expected to expand further in Saudi Arabia, reflecting Saudi government initiatives on reducing spending due to weak oil prices, and because of the benefits of potential cost savings [85]. Figure 3 shows the total market for ICT and Cloud Computing in Saudi Arabia over three years [87, 88].

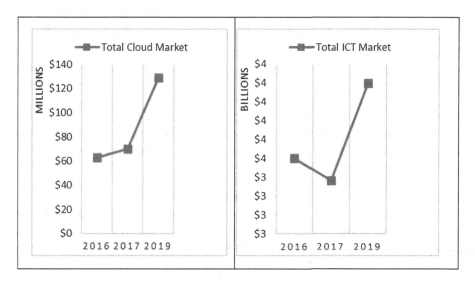

Fig. 3. Total Market for ICT and Cloud Computing in Saudi Arabia for four years (adapted by author). Source: [87, 88].

The Saudi government established the Vision 2030 initiative to support migration from an oil-based economy to a non-oil economy [103]. The National Transformation Program 2020 (NTP) was announced as a part of this initiative [89]. The NTP aims to support Digital Transformation Initiatives and improve the effectiveness of public sector organisations [90]. ICT technologies can play important roles in facilitating this trans-formation by supporting innovative solutions that improve efficiency at lower costs [91]. Cloud Computing solutions particularly can provide support for the government initia-tives by enhancing cross-government collaborations and providing the other benefits of Cloud Computing. However, Saudi Arabia has still not undertaken a national Cloud Computing initiative that could support the growth of Cloud Computing.

5.2 Studying Cloud Computing Adoption in Saudi Arabia

Cloud Computing in Saudi Arabia started to receive attention from 2013 [92]; before then, little research had been conducted on the implementation of Cloud Computing in the country. Alharbi [93] and Alotaibi [94] studied users' acceptance of Cloud Computing in Saudi Arabia based on the Technology Acceptance Model (TAM). Although the studies provided insights into the factors affecting Cloud Computing adoption in Saudi Arabia, both studies implemented TAM to predict users' acceptance of Cloud Computing. From an organisation level, Yamin (2013) completed a survey of Cloud Computing awareness in Saudi Arabia, showing that Cloud technologies represent a new trend for Saudi organisations. However, this research provided only a general view of Cloud Computing adoption in Saudi Arabia. Alkhater et al. [95] investigated influential factors in the decision to adopt Cloud Computing in general. They indicated that many factors such as trust, relative advantage and technology readiness will influence the use of Cloud Computing technology. However, they did not investigate the effect of two dimensions – human and business – on the implementation of Cloud Computing. Another limitation of their study is the sample size, which was small (i.e. only 20 experts). Alsanea [96] investigated the adoption of Cloud Computing in the government sector in general in Saudi Arabia. Their study indicated that there is a high possibility of Cloud Computing acceptance among Saudi government organisations since 86% of the participants supported the adoption of Cloud Computing in their organisations. The study also showed that Cloud Computing adoption in governmental sectors is positively affected by indirect benefits of the cloud, industry type, cost, trust and feasibility. However, the study provides views of the factors affecting Cloud Computing implementation in Saudi government organisations without providing specific information about the type of organisation, and it did not include human factors.

Alhammadi et al. [97] studied the determinants of Cloud Computing adoption in Saudi Arabia and found that factors influencing the decision whether to adopt Cloud Computing are security concerns, organisation readiness, top management support, firm status, government support and compatibility. This study was at organisation level and did not focus on a specific industry and did not provide information about the participants' roles. Alkhlewi et al. [98] identified 15 factors for the successful implementation of a Private Government Cloud in Saudi Arabia. However, the study focused only on government organisations and the sample size was small (i.e. only 30 experts) and the participants were mainly IT professionals. Albar and Hoque [99] proposed a theoretical framework to study Cloud ERP Adoption in Saudi Arabia without applying the framework.

Tashkandi and Al-Jabri [100] studied Cloud Computing adoption by higher education organisations in Saudi Arabia. The study focused on: technological, organisational and environmental factors. They found that relative advantage has a positive influence on the decision on Cloud Computing adoption. However, complexity and vendor lock-in were found to have a negative impact. The study has some limitations since it is mainly focused on higher education organisations and the researchers do not include the business dimension, which is an important dimension in the adoption of Cloud Computing. Noor [101] examined the usage of Cloud Computing in Saudi universities; 300 participants from five different universities participated in the study. The study found that the key drivers of university IT department employees' Cloud adoption were the ability to access the Cloud via any device and the self-service feature of Cloud Computing. It also indicated that the main barriers to the adoption of Cloud Computing were privacy, compliance, security, reliability and availability respectively. The study focused mainly on Saudi universities at the user level and ignored other aspects such as cost of adoption. Almutairi and El Rahman [102] studied the impact of IBM Cloud Solutions on Saudi students and showed that the majority of the students knew about Cloud Computing;

however, only 56% of the participants had used Cloud solutions. Albugmi et al. [103] proposed a theoretical framework for Cloud Computing adoption by Saudi government overseas agencies without testing the framework against specific respondents or case studies. Similarly, Mreea et al. [104] proposed a Cloud Computing value model for public sector organisations in Saudi Arabia without testing the framework against specific respondents or case studies. Alharthi et al. [105] investigated critical success factors for Cloud migration in Saudi universities and identified six factors: security, reliability, interoperability, migration plan, regulation compliance and technical support with the Arabic language. The paper did not discuss business factors such as cost and technical factors such as infrastructure readiness.

5.3   Cloud Computing in the Saudi Healthcare Context

In the Saudi healthcare domain, few studies have discussed the use of Cloud Computing in Saudi healthcare organisations. Cloud Computing has been implemented by King Abdulaziz City for Science and Technology (KACST) to support the large Saudi Genome Project [106]. Azzedin et al. [107] developed the Disease Outbreak Notification System (DONS) in Saudi Arabia. However, both studies only considered the technical implementation of Cloud Computing and did not study the factors influencing Cloud Computing adoption in Saudi healthcare organisations.

Cloud Computing may assist in solving some of the management challenges of healthcare organisations in Saudi Arabia. Since financial issues are affecting e-health projects in the country, Cloud Computing can offer economic savings by decreasing the initial and operational costs of e-health projects in Saudi hospitals. Cloud Computing could help to reduce the problem of shortage of IT and health informatics technicians since the use of Cloud technology means that fewer technicians will be required by the healthcare organisations [17]. Cloud-based medical applications will also enable IT departments in healthcare organisations to focus more on supporting the implementation of e-health projects by moving some of their responsibilities to the Cloud providers' side, particularly in a public Cloud Computing environment. For healthcare organisations, Cloud Computing will enable better integration and exchange of medical records across multiple organisations [1]. Using Cloud Computing in Saudi healthcare organisations will facilitate the provision of sufficient computing resources to deal with the large amount of data that is created by e-health services. This feature will also help Research and Development (R&D) departments in healthcare organisations at the national level [1]. Cloud Computing used in collaboration with other technologies such as the Internet of Things, m-health and Big Data will help reshape healthcare services in Saudi Arabia. Cloud Computing solutions will be suitable technologies to support demands on Saudi healthcare since the country's population is expected to increase from 30m to 37m by 2030 [108]. Cloud Computing technology will allow Saudi healthcare organisations to enhance their information-processing capacity by sharing IT resources which include software, hardware and expert skill sets. Cloud Computing could help in solving the fragmentation and isolation problems of the Saudi healthcare information system. Cloud Computing's elasticity feature, which is the ability to scale the IT services dynamically and quickly, could be appropriate for Saudi healthcare demands [4]. A potential use of this feature could be during Hajj session when Saudi Arabia hosts from two to three million people for a specific time (i.e. one to three months) every year. Cloud Computing in collaboration with other technologies such as the Internet of Things, m-health and Big data could help reshape healthcare services in Saudi Arabia. To support the successful adoption of Cloud Computing in the Saudi Healthcare environment, this study investigates the factors which influence health care decision makers when deciding whether to adopt Cloud Computing.

6   Research Methodology

The findings of the previous sections identified that Cloud Computing implementation in Saudi healthcare organisations requires further investigation. Thus, the aim of the research is to develop an in depth understanding of the factors which influence Cloud Computing adoption in Saudi Healthcare organisations.

Due to the exploratory nature of this research, a case study approach was considered appropriate. Case studies are useful for exploring phenomena where existing knowledge is limited and support examination in depth of a complex phenomenon in its natural setting [2]. A multiple-case study methodology was selected to carry out this study by choosing two hospitals in two different cities in Saudi Arabia. One hospital was based in the capital of Saudi Arabia and another at a major city in Saudi Arabia. The hospitals were at different stages of the Cloud Computing adoption process. Due to confidentiality requirements, data has been anonymised.

The focus of this study was on the Cloud Computing decision-making process and the people who could affect the decision. Although this study considered IT department mangers as the main decision makers for the implementation of Cloud Computing, other decision makers such as hospital directors and medical directors were also included to get the views of other stakeholders. The selected participants were in charge of the adoption and selection of information systems solutions in the selected hospitals. Seven senior decision-makers drawn from the two hospitals agreed to participate in the study. The participant responses are coded from P1 to P7 but to preserve confidentiality, details of the individual's roles are not given. The primary data collection method was semi-structured qualitative interviews. Interviews were used because this allows the researchers to interact directly with the participant and supports clarification of concepts [2]. The interview method also supports understanding by meeting the respondents in their social context [40]. While interviews were conducted in person face-to-face meetings, all respondents were assured anonymity to promote openness. The thematic analysis approach was applied to allow the researchers to report the findings of the study. This approach is used in qualitative research to identify, analyse and report patterns (themes) within data [109]. Each theme represents something important about the data in relation to the research activities [110]. By classifying the data into categories through comparison between and within cases, the themes are developed and realised [109]. A framework analysis method which is one method of thematic analysis was applied in the analysis of the data in this study. In framework analysis, the data is summarised by using matrix output where rows represent cases and columns represent codes [109]. We conducted the framework analysis by applying the steps mentioned in [110] as follow:

- Familiarisation with the data: This was achieved through the process of transcribing interviews and reading the transcripts many times and the participants checked the accuracy of the information.
- Generating codes: In this step, the transcripts are read to generate an initial list of ideas about elements which appear interesting. This step is helpful to ensure important aspects of the data are not missed [109].
- Allocating themes: data was re-analysed to group the codes together into categories or themes. Iteration may be required at this stage to review, define and name the themes [109]. In this step, the researchers identified that factors affecting Cloud Computing adoption in Saudi Healthcare organisations could be divided into five main themes which are Technology, Organisation, Environment, Human and Business.
- Producing the report: the researchers produced the final report showing the findings and the full thematic analysis.

7   Results

The following sections present the findings of the semi structured interviews with the decision makers in Saudi healthcare organisations.

7.1   Cloud Computing Drivers

The study identified motivations for the adoption of Cloud Computing in Saudi healthcare organisations. The key benefits of adopting Cloud Computing in Saudi healthcare organisations, found in this study, are as follows:

- **Reducing IT cost:** Cost saving is considered to be one of the major motivations for using Cloud Computing services. This driver is important for many participants (P2, P5, P6). One participant stated that "*Cost saving is the number one factor that affected our decision when adopting Cloud Computing.*" (P5). The impact of cost savings will be either direct or indirect. Direct cost savings can be caused by decreasing the need for capital investment for IT resources (P5). Indirect cost savings include decreasing the cost of power consumption (P7) and decreasing the maintenance cost (P3, P6). The use of Cloud Computing solutions will reduce the need for new IT resources for development and testing of new software (P3).
- **Collaboration between healthcare organisations:** Cloud Computing will enable better integration and exchange of medical information across multiple organisations (P1, P4). One participant stated that they had a proposal to connect all the hospitals in the region via Community Cloud Computing (P4).
- **Increase resources sharing:** Some healthcare organisations have shortages issues in terms of advanced medical devices. Some participants see Cloud Computing as an enabling technology that allows them to share their resources (P1, P4, P5). One

participant stated that *"The use of Cloud Computing could allow us to share some of medical devices with other organisations such as: PACS."* (P1).

- **Better IT resources Manageability:** Maintenance and operation activities are time consuming for IT employees. Cloud Computing will allow for better management of these activities by decreasing the number of IT employees required for these activities. IT resources can be acquired and deployed more quickly and easily due to the self-service characteristic of cloud computing. For example, one participant explained that *"The hospital has many buildings which require several minutes for IT staff to go between buildings. Thus, Cloud Computing will reduce maintenance time since some maintenance jobs will be done by using virtualisation technologies."* (P3).

- **Address the issue of scarcity of professional:** some Saudi healthcare organisations face a shortage of professionals either as IT professionals or health informatics professionals [2]. Some participants pointed out that Cloud Computing could help to reduce the problem of the shortage of IT and health informatics technicians since the use of Cloud technology means that fewer technicians will be required by the healthcare organisations (P1, P2).

- **New business opportunities:** The use of Cloud computing in healthcare organisations will also allow the introduction of new classes of applications and delivery of services that were not possible before [4]. Comments made in the interviews suggest that Cloud Computing can help healthcare organisations with the introduction of new applications that were not previously available because of cost or technical difficulties (P1). Another advantage of implementing Cloud Computing solutions is that Cloud Computing will give IT professionals more time to think about innovative ides because it will reduce daily operations and maintenance time (P7).

- **Enhance Business Agility:** Cloud computing can allow healthcare organisations to add to or change its IT resources quickly and easily without the need to wait a long time for replacement (P3, P5). One participant clarified that by saying *"We will not need to wait a long time for someone to accept our request to add new hardware or software."* (P3).

- **Improve IT resources availability:** Cloud Computing will allow healthcare organisations to improve the availability of their IT resources (P3, P4, P6). For example, one participant stated that *"Another benefit of using Cloud Computing is that we will not need to have application down time when we conduct planned maintenance."* (P6). Cloud Computing will also support development and testing of new components of the system with limited need for extra IT resources (P3).

- **Scalability:** This driver refers to the ability of IT resources to scale up and down based on the needs of the users' demands and usage [1]. Only one participant mentioned scalability as a driver of Cloud Computing (P7). However, scalability may not be an issue for Saudi Healthcare organisations which usually have unchanging usage of IT resources [2].

Table 2 summarises the key drivers for Cloud Computing adoption in Saudi healthcare organisations based on the participants' views.

Table 2. Key drivers for cloud computing adoption in Saudi healthcare organisations

Key driver	Participant code
Cost reduction	P2, P3, P5, P6
Collaboration with other organisations	P1, P4
Resource sharing	P1, P4, P5
Better IT resources Manageability	P3
Reduce the need for qualified IT staff	P1, P2
New business opportunities	P1, P6, P7
Enhance business Agility	P3, P5
Improve the availability of IT resources	P3, P4, P6
Decrease power consumption	P6
Scalability	P7

7.2 Cloud Computing Challenges

The participants of this study also discussed the challenges that may hinder the adoption of Cloud Computing in Saudi Healthcare organisations as follow:

- **Hidden cost:** Although the cost of Cloud Computing model is considered to be more transparent than traditional outsourcing [111], some participants have concerns about the hidden cost of implementing this model (P1, P3). One possible reason for this is the lack of experience of dealing with such a model (P1). A solution suggested during the interviews was to conduct a complete financial analysis before adopting Cloud Computing.
- **Security concerns:** Healthcare organisations usually have concerns regarding IT security because of the sensitive personal data that they deal with [1]. Security issues have been identified as a challenge for adopting Cloud Computing solutions by some participants (P1, P5). However, one participant stated that "*I am not concerned that much about security issues because I hope that we at the hospital can deal with such concerns. We can deal with security concerns through technological means such as use of firewall.*" (P2). One possible explanation for the security concerns is the shortage of professionals in Saudi Arabia who can deal with such concerns [2]. The implementation of Private Cloud solutions could minimise these concerns.
- **Data Privacy concerns:** the privacy and confidentiality of hospital information is a real concern for some participants (P3, P6). The concerns are about who has access to the data and the fear of losing the control over the data. However, one participant indicated the availability of national regulation about Cloud Computing will help to ease some of these concerns (P5). Patient information privacy is a concern for Saudi hospitals since the country does not have national health privacy legislation [2].
- **Integration and compatibility concerns:** One participant clarified that some software at the hospital is found to be incompatible with Cloud Computing services (P5). This will lead to difficulties in the integration processes between some IT resources and Cloud Computing environment (P4). One possible reason for this obstacle is the shortage of qualified IT professionals at the hospital who are able to make any

required modifications for the software or the hardware to be compatible with Cloud Computing.

- **Managing the relationship with Cloud Vendor:** healthcare organisations will need to manage the relationship with their Cloud provider very carefully (P7). Some healthcare organisations have concerns about Vendor-lock in (P4). However, participants identified possible solutions for such concerns. One participant stated that *"The hospital must have a database migration strategy to avoid vendor lock-in issue"* (P4). Another participant mentioned that standardisation of Cloud Computing could solve this issue.

Table 3 summarises the challenges for Cloud Computing adoption in Saudi healthcare organisations from decision makers' point- views.

Table 3. Challenges for cloud computing adoption in Saudi healthcare organisations

Challenge	Participant code
Hidden cost	P1, P3
Security concerns	P1, P2, P5
Data privacy concerns	P3, P5, P6
Integration and compatibility concerns	P4, P5
Managing relationship with the vendor	P4, P7

7.3   Factors Affecting Decision Makers at Healthcare Organisations to Adopt Cloud Computing

The factors that affect the decision making process of adopting Cloud Computing at Saudi healthcare organisations based on the participants' views, are classified as follows:

- **Technological Factors**

The awareness among healthcare organisations about the potential benefits of adopting Cloud Computing is seen as a positive factor supporting the adoption of cloud computing. Relative advantage was considered by all respondents since they stated that their organisations are aware of the values of implementing Cloud Computing solutions. Another factor mentioned by all participants is the IT infrastructure readiness of the healthcare organisation to implement Cloud Computing. Some participants were concerned about Internet connectivity at their organisation (P1, P5, P6), while other participants were concerned about integrating Cloud Computing with legacy architectures (P2, P4, P6). Compatibility was cited by many respondents as an important factor that will affect the decision about Cloud Computing adoption. This factor was discussed from two perspectives which are IT infrastructure compatibility with the Cloud (P3, P5) and healthcare organisations values compatibility (P1). Security concerns are also considered to have an impact on Cloud Computing implementation at healthcare organisations (P1, P4, P5). However, the implementation of Private Cloud solutions could ease some security concerns. One participant mentioned that providing trialability (i.e. the degree to which an innovation can be experimented prior to actual adoption [112]) of using Cloud Computing solutions will affect positively the decision of Cloud

Computing (P7). However, this trialability factor will be dependent on the innovativeness of IT leaders at the organisations.

- **Organisational Factors**

Top management support was identified by almost all the respondents as an important factor for Cloud Computing adoption. However, some of them indicated that this factor has affected the adoption decision positively (P1, P5) and others stated that top management had negatively affected the decision (P2, P3, P6, P7). One participant suggested that lack of knowledge about Cloud Computing maybe the reason for lack of top management support. Another factor mentioned by some of the participants was the attitude toward change, mostly from IT personnel at the healthcare organisations. The data indicated that there is a possibility for resistance to change especially from IT staff at the organisations (P1, P2, P3, P6, P7). One possible reason for the resistance is the fear of IT staff that they may lose their jobs due to the adoption of Cloud Computing. One participant stated that explaining Cloud Computing concepts for all stakeholders of the hospital would be important (P2). However, some interviewers did not believe that Cloud Computing adoption will affect end users nor IT staff (P4, P5).

- **Environmental Factors**

The data revealed that national regulation about the implementation of Cloud Computing can play an important role in supporting or slowing the adoption of Cloud Computing (P3, P4, P5, P6). Currently, there is no national regulation about Cloud Computing in Saudi Arabia. However, some participants indicated that there are general IT regulations which will apply to the adoption of Cloud Computing in healthcare organisations such as: the resolution of Computing and Networking Controls in Government Agencies in Saudi Arabia (P4). Pressures from peers in the same industry will also affect Cloud Computing adoption. Almost all of the respondents agreed that if pioneering hospitals in Saudi Arabia adopted Cloud Computing, this will influence other hospitals to adopt it quickly. Some participants clarified that visiting hospitals that had already implemented Cloud Computing solutions would allow them to learn from their experience and would make the implementation process easier. The availability of external support for implementation and use of Cloud Computing will also be a positive factor supporting adoption. Healthcare organisations in Saudi Arabia will need to work with IT providers that have a good reputation in the area and have multi-disciplinary experience (i.e. both health informatics and IT) (P5). Another participant stated that *"attending seminars organised by Cloud providers affected their decision about adopting Cloud Computing"* (P7).

- **Human Factors**

The innovativeness of the decision-makers had a great impact on the willingness to adopt Cloud Computing in Saudi healthcare organisations (P1, P3, P4, P6, P7). Chief Information Officer (CIO) innovativeness allowed the hospital to try the latest technologies and to find innovative ways of implementing IT services at the hospital (P4, P6, P7). CIOs can use their capabilities to convince top management to adopt new technologies (P1). The availability of internal experts is another important factor that should be considered when making the decision about Cloud Computing adoption in healthcare organisations (P2, P3, P5, P6, P7). However, some participants indicated that the roles of IT staff will be changed due to the implementation of Cloud Computing to focus more on managerial activities instead of operational activities (P2) which could be related to possible resistance to change from IT staff. Another respondent mentioned that the hospital may need to recruit IT staff that are able to deal and manage Cloud Computing solutions effectively and it could be expensive to recruit them. Additionally, previous experience of implementing and managing this kind of project is considered another factor by many participants (P1, P3, P4, P5, P6). Many participants advised the organisations to have a training plan for their IT staff to develop knowledge and experience (P2, P6, P7). One participant mentioned that *"attracting new IT professionals with creative minds and providing them with enough training about Cloud Computing"* could be a solution (P2).

- **Business Factors**

There was agreement among all participants in this study that healthcare organisations must consider financial aspects before adopting Cloud Computing. Healthcare organisations must be aware of the direct costs (e.g. new hardware) and non-direct costs (e.g. training) of implementing Cloud Computing. One participant clarified that establishing detailed Service Level agreements (SLA) is important for the organisation before Cloud Computing is adopted (P5). The participants also highlighted that an important factor to be considered when adopting Cloud Computing is the strategic benefits of the adoption. Some participants indicated that Cloud Computing will increase collaboration between the hospital and other organisations (P1, P4). One participant stated that *"We are implementing new technologies but with old processes so I think that the business model of our department needs to be changed"* (P7).

8 Discussion

The goal of this study is to assess the determinants of cloud computing adoption in Saudi healthcare organisations by means of interviews with decision makers in Saudi Healthcare organisations. This study identified five groups of factors which will influence Cloud Computing adoption in Saudi Healthcare, shown in Table 4.

The next sections discuss the findings of the interviews and provide some recommendations for Saudi healthcare organisations which are planning to adopt Cloud Computing solutions.

Table 4. Factors which affect cloud computing adoption decision in Saudi healthcare organisations

Perspective	Sub-factor	Participant code
Technology	Internet connection	P1, P5, P6
	Compatibility	P1, P3, P5
	Relative advantages	P1, P2, P3, P4, P5, P6,P7
	Integration	P2, P6, P4
	infrastructure readiness	P2, P3, P4, P6, P7
	Security	P1, P4, P5
	Trialability	P6
Organisation	Top management support	P1, P2, P3, P5, P6, P7
	Change resistance	P1, P2, P3, P6, P7
	Lack of knowledge of Cloud Computing	P3
Environment	Regulation compliance	P1, P3, P4, P5
	Business ecosystem pressure	P1, P2, P3, P5, P6, P7
	External expertise	P1, P3, P4, P7
	Lack of standardisation	P2, P4
	Accreditation organisation	P5
Human	Innovativeness	P1, P3, P4, P6, P7
	Internal expertise	P2, P3, P5, P6, P7
	Previous experience	P1, P3, P4, P5, P6
Business	Cost and hard financial analysis	P1, P2, P3, P4, P5, P6
	Strategy and soft financial analysis	P1, P2, P3, P4, P5, P6, P7

8.1 Technology

Technological factors describe the technologies available to an organisation and how they will influence the Cloud Computing adoption process [113]. Four technical factors are identified in the current study which are relative advantage, technological readiness, compatibility and security. Before making any decision about the adoption of an innovation, decision makers usually start by discovering the relative advantages of such innovation [113]. The relative advantage of Cloud Computing is the degree to which Cloud Computing solutions have clear benefits over others in meeting organisation's needs [80]. Respondents in this study agreed that Cloud Computing will provide many benefits for healthcare organisations. These benefits include improving business agility, improving the availability of IT resources, cost savings, improving collaboration between healthcare organisations. Thus, relative advantage will be a positive factor on the decision to adopt Cloud Computing in Saudi healthcare organisations. This finding is consistent with other studies that found Cloud Computing can provide many benefits for the organisation [12, 40]. However, some studies either found relative advantage to have a negative impact on Cloud Computing adoption in the high-tech industry [113] or to be unimportant as in Taiwan hospitals [80]. The finding in the current study is consistent with other studies that found perceived benefits have positive impact on the Cloud adoption decision [25]. A factor mentioned by all the participants in the current study is technology readiness which refers to technological infrastructure at the health-care organisations such as network technologies and IT systems [113]. Participants have concerns about the ability of their IT infrastructure to cope with some of the technical characteristics of Cloud Computing. This finding is similar to the findings by HIMSS Analytics Cloud Survey which indicated that in order to use Cloud Computing services, healthcare organizations have to upgrade their network infrastructure [114]. Our finding implies that healthcare organisations in Saudi Arabia will need to upgrade some of their IT infrastructure in order to implement cloud solutions. Example of required upgrades mentioned by the participants is network bandwidth.

Other studies have shown that technology readiness is a facilitator of Cloud Computing adoption [12]. Healthcare organisations in Saudi Arabia also have some concerns about the compatibility of Cloud Computing either with current IT infrastruc-ture or healthcare values. One possible explanation for these concerns is the shortage of qualified IT professionals at the hospital able to deal with the integration processes. Although, compatibility has been found not to have a significant effect on Cloud Computing adoption in Saudi academic organisations [100], the situation in healthcare organisations is different. The findings of our study are consistent with the findings of [115] who indicated that issues with compatibility had a negative impact on Cloud Computing adoption at US manufacturing and retail industries. Other researchers found that compatibility is a facilitator for Cloud Computing adoption decision in the services sector in Portugal [12].

Security and privacy are also concerns for Saudi healthcare organisations when adopting Cloud Computing, given the sensitive nature of health information. Regarding privacy, some respondents referred to regulatory compliance. This finding is consistent with other studies where data security was found to be the most critical factor affecting the decision to adopt cloud computing in Taiwanese hospitals [80]. However, security concerns were not identified as hindering the adoption of Cloud Computing in Portu-guese firms [12]. Privacy and data security may be more significant issues in a health context. The IT management team should also stipulate that the Cloud Computing vendors must have an effective disaster recovery plan ready in case of any system failure issues.

8.2 Human

Human factors refer to the internal capabilities of the organisation's workforce and the necessary skills and technical competence [12]. Although some participants indicated that Cloud Computing could address issues of scarcity of IT professionals, others mentioned the shortage of qualified IT staff able to deal with cloud technology and that recruitment would be expensive. The dependence of Saudi healthcare organisations on trading partners for their IT solutions could offer an explanation for this result [2]. The availability of internal Information Systems (IS) expertise was found to affect the deci-sion on adoption of Cloud Computing in Taiwanese hospitals [80]. Our study also indi-cated that the roles of hospital IT staff may require some modifications. As Cloud

Computing is an emerging technology, prior technology experience is important for Saudi healthcare organisations. However, since the use of Cloud Computing is still developing in the healthcare field, the respondents identified that more training will be required for IT staff in Saudi healthcare organisations. Lack of knowledge of Cloud Computing was found to have a negative impact on Cloud Computing adoption [5]. CIO innovativeness and capabilities are found to have positive impact on the decision on implementing Cloud Computing in healthcare in Saudi Arabia. The participants indicated that CIOs could use their networks to influence top management to support Cloud Computing adoption. Hospitals usually are slow in adopting new information technologies and decision makers at hospitals prefer to use mature technologies for patients safety reasons [116]. It could be important for Saudi healthcare organisations to consider updating IT staff skills and knowledge to provide for more effective management of IT human resources, and consider building a human resources skills inventory database.

8.3 Organisation

Organisational characteristics of healthcare organisations will play an important role in decision making about the implementation of Cloud Computing [12]. These characteristics are the organisational conditions that enable or limit the adoption and implementation of Cloud Computing [80]. The participants in this study identified that two factors from the organisation perspective which affect Cloud Computing adoption in Saudi healthcare are top management support and attitude toward change. With regard to top management support, the participants in this study did not agree whether the impact was positive or negative. This confusion may refer to the fact that decisions about Cloud Computing will require approval from different levels of management (i.e. hospital directors and Ministry of Health directors). One possible reason for a perceived negative impact of top management may be lack of knowledge about Cloud Computing as mentioned by some participants. Another possible reason is that some managers may still have some concerns about the implementation of Cloud Computing in healthcare organisations [100]. However, some participants indicated that top management support was a positive factor in the adoption of Cloud Computing. One possible explanation is that hospital mangers have identified the benefits of adopting Cloud Computing in their organisations [80]. Another important factor in the organisational perspective is attitude toward change. This factor is important because the adoption of Cloud Computing will affect the whole organisation not only specific units or departments [113]. This factor was found to have a negative impact on Cloud Computing adoption [114] which is consistent with the finding of our study. One possible explanation for this is the fear of IT staff that they may lose their jobs due to the adoption of Cloud Computing. Participants advised healthcare organisations to organise workshops that explained the Cloud Computing concept to all the stakeholders in the organisation. The results identified some positive attitude towards change in relation to Cloud Computing at the hospitals. However, a Change Management Team and the processes involved will help alleviate the negative concerns about the implementation of Cloud Computing in healthcare organisations.

8.4 Environment

Environmental factors refer to the healthcare environment and interactions with the government [12]. Regarding regulatory compliance, there is currently no national regulation for Cloud Computing in Saudi Arabia. However, the participants clarified that the implementation of Cloud Computing must comply with all IT regulations in Saudi Arabia. Other researchers [100, 117] found that regulatory compliance is not a determinant of Cloud Computing adoption. That could be because compliance is mandatory for healthcare organizations so they do not need to consider it when making the decision about Cloud Computing [35]. Another, environmental factor is peer pressures in the healthcare environment. This factor was found to have a positive impact on the adoption of Cloud Computing since almost all the respondents indicated that the implementation of Cloud Computing at other hospitals will affect their decision positively. This finding is similar to the findings of other studies that found external pressure to be positively related with Cloud Computing adoption [100, 117]. Another related factor is the availability of IT vendors that can provide successful implementation of Cloud Computing for healthcare organisations. This factor was found to affect Cloud Computing adoption in Saudi healthcare organisations positively. This finding is consistent with other studies that found the availability of IT providers with good capabilities of support and reputation to be important [5, 117]. However, some participants indicated that healthcare organisations must be able to manage the relationship with IT vendors effectively.

8.5 Business

Healthcare organisations are also required to analyse the implementation of Cloud Computing from the business side. This should take into consideration quantitative metrics (i.e. cost analysis) and qualitative metrics (i.e. intangible aspects of Cloud Computing adoption decision) [61]. Our study found that cost saving is one of the main drivers of Cloud Computing adoption. This finding is supported by other studies that found cost saving to be an important factor when deciding to adopt Cloud Computing [12, 80]. However, there are still some concerns about the hidden cost of implementing this technology. Participants advised healthcare organisations to pay great attention to defining their requirements to avoid unnecessary costs. Another positive factor influencing the adoption of Cloud Computing in Saudi healthcare organisations is soft financial analysis which considers the business opportunities and benefits that Cloud Computing provides for the organisations. Participants listed some benefits such as new business opportunities and improving collaboration between healthcare organisations. Thus, organisations must also consider the intangible aspects of adopting Cloud Computing solutions. However, some participants emphasised the possible change for the business model of IT departments at the organisations when implementing Cloud Computing solutions.

9 Conclusion

Cloud Computing can offer many benefits for healthcare organisations such as cost savings, patient care improvement, research support and overcoming the issues of resource scarcity. However, the implementation of Cloud Computing may face some issues such as concerns regarding to security and data privacy, issues of availability and reliability of Cloud services, hidden cost. The study critically reviewed a number of Cloud Computing adoption models and the status of Cloud Computing in Saudi Arabia. This study aimed to provide understanding of Cloud Computing adoption decision-making in healthcare organisations in Saudi Arabia to make clear decisions about Cloud Computing adoption. This investigation applied semi structure interviews with decision makers in Saudi healthcare organisations to gain an in-depth understanding of Cloud Computing in the Saudi context. This study indicated that five perspectives, Business, Technology, Organisational, Environmental and Human, should be considered when deciding whether to adopt Cloud Computing. The results of the interviews carried out with health sector decision makers showed that the factors which determine Cloud Computing adoption in Saudi healthcare organisations are relative advantage, technology readiness, compatibility, security, innovativeness, internal expertise, previous experience, hard financial analysis, soft financial analysis, regulation compliance, business ecosystem partner pressure, external expertise, top management support and attitude towards change. Future work is to develop a strategic decision-making framework for Cloud Computing adoption, based on the factors identified in this study, to support a holistic approach to Cloud Computing Decision Making.

References

1. AbuKhousa, E., Mohamed, N., Al-Jaroodi, J.: E-health cloud: opportunities and challenges. Futur. Internet **4**(4), 621–645 (2012)
2. Alkraiji, A., Jackson, T., Murray, I.: Barriers to the widespread adoption of health data standards: an exploratory qualitative study in tertiary healthcare organizations in Saudi Arabia. J. Med. Syst. **37**(2), 9895 (2013)
3. Calabrese, B., Cannataro, M.: Cloud computing in healthcare and biomedicine. Scalable Comput. Pract. Exp. **16**(1), 1–18 (2015)
4. Marston, S., Li, Z., Bandyopadhyay, S., Zhang, J., Ghalsasi, A.: Cloud computing — the business perspective. Decis. Support Syst. **51**(1), 176–189 (2011)
5. Güner, E.O., Sneiders, E.: Cloud computing adoption factors in Turkish. In: PACIS 2014 Proceedings, p. 353 (2014)
6. El-Gazzar, R., Hustad, E., Olsen, D.H.: Understanding cloud computing adoption issues: a Delphi study approach. J. Syst. Softw. **118**, 64–84 (2016)
7. Voas, J., Zhang, J.: Cloud computing: new wine or just a new bottle? IT Prof. **11**(2), 15–17 (2009)
8. Mell, P., Grance, T.: The NIST definition of cloud computing. National Institute of Standards and Technology (U.S. Department of Commerce), USA (2011)
9. North Bridge. Sixth Annual Future of Cloud Computing Survey (2016)
10. IDG Enterprise. Cloud Computing Survey (2016)
11. Cloud Industry Forum. UK Cloud adoption snapshot & trends for 2016 (2016)

12. Oliveira, T., Thomas, M., Espadanal, M.: Assessing the determinants of cloud computing adoption: an analysis of the manufacturing and services sectors. Inf. Manag. **51**(5), 497–510 (2014)
13. Harvard Business Review. Cloud: Driving a Faster, More Connected Business (2015)
14. Durao, F., Carvalho, J.F.S., Fonseka, A., Garcia, V.C.: A systematic review on cloud computing. J. Supercomput. **68**(3), 1321–1346 (2014)
15. CFO. Defining the Business Value of Cloud Computing (2012)
16. Weinman, J.: Cloudonomics: The Business Value of Cloud Computing. Wiley, New York (2012)
17. Sultan, N.: Making use of cloud computing for healthcare provision: opportunities and challenges. Int. J. Inf. Manage. **34**(2), 177–184 (2014)
18. Garg, S.K., Buyya, R.: Green cloud computing and environmental sustainability. In: San, M., Gangadharan, G.R. (eds.) Harnessing Green IT: Principles and Practices, pp. 315–340. Wiley, New York (2012)
19. Carroll, M., van der Merwe, A., Kotze, P.: Secure cloud computing: benefits, risks and controls. Information Security South Africa (ISSA), pp. 1–9 (2011)
20. Armbrust, M., Fox, A., Griffith, R., Joseph, A.D., Katz, R., Konwinski, A., Lee, G., Patterson, D., Rabkin, A., Stoica, I.: A view of cloud computing. Commun. ACM **53**(4), 50–58 (2010)
21. RightScale. State of the Cloud Report (2015)
22. Thakur, A., Kumar, M., Yadav, R.K.: Cloud computing: demand and supply in banking, healthcare and agriculture. J. Comput. Eng. **16**(3), 96–101 (2014)
23. Harvard Business Review and Oracle. Cloud Computing Comes of Age (2015)
24. Nicholson, B., Owrak, A., Daly, L.: Cloud Computing Research (2013)
25. Hsu, P.-F., Ray, S., Li-Hsieh, Y.-Y.: Examining cloud computing adoption intention, pricing mechanism, and deployment model. Int. J. Inf. Manage. **34**(4), 474–488 (2014)
26. Morgan, L., Kieran, C.: Factors affecting the adoption of cloud computing: an exploratory study. In: The 21st European Conference on Information Systems, pp. 1–12 (2012)
27. Chen, Y., Ku, W.-S., Feng, J., Liu, P., Su, Z.: Secure distributed data storage in cloud computing. In: Cloud Computing: Principles and Paradigms. Wiley (2011)
28. Phaphoom, N., Wang, X., Samuel, S., Helmer, S., Abrahamsson, P.: A survey study on major technical barriers affecting the decision to adopt cloud services. J. Syst. Softw. **103**, 167–181 (2015)
29. North Bridge. Future of Cloud Computing 2012 (2012)
30. Oxford Economics and SAP. The new engine of business (2015)
31. Avižienis, A., Laprie, J.C., Randell, B., Landwehr, C.: Basic concepts and taxonomy of dependable and secure computing. IEEE Trans. Depend. Sec. Comput. **1**(1), 11–33 (2004)
32. Mather, T., Kumaraswamy, S., Latif, S.: Cloud Security and Privacy: An Enterprise Perspective on Risks and Compliance. O'Reilly Media Inc., Sebastopol (2009)
33. Cloud Security Alliance. Cloud Computing Top Threats in 2016 The Treacherous 12 (2016)
34. KPMG. 2014 Cloud Survey Report: Elevating business in the cloud (2014)
35. Borgman, H.P., Bahli, B., Heier, H., Schewski, F.: Cloudrise: exploring cloud computing adoption and governance with the TOE framework. In: 2013 46th Hawaii International Conference on System Science, pp. 4425–4435, January 2013
36. Kar, A.K., Rakshit, A.: Flexible pricing models for cloud computing based on group decision making under consensus. Glob. J. Flex. Syst. Manag. **16**, 191–204 (2015)
37. Right Scale. State of the Cloud Report (2016)
38. Chung, D.: Adoption and implementation of cloud computing services : a railroad company case. Issues Inf. Syst. **15**(II), 276–284 (2014)

39. Toosi, A.N., Calheiros, R.N., Buyya, R.: Interconnected cloud computing environments. ACM Comput. Surv. **47**(1), 1–47 (2014)
40. Lin, A., Chen, N.: Cloud computing as an innovation: perception, attitude, and adoption. Int. J. Inf. Manage. **32**(6), 533–540 (2012)
41. Desai, N.S.: Mobile cloud computing in business. Int. J. Inf. Sci. Tech. **6**(1/2), 197–202 (2016)
42. Khajeh-Hosseini, A.: Supporting System Deployment Decisions in Public Clouds. University of St Andrews (2012)
43. Alhammadi, A.: A Knowledge Management Based Cloud Computing Adoption Decision Making Framework. Staffordshire University (2016)
44. Alkhalil, A.: Migration to Cloud Computing: A Decision Process Model. Bournemouth University (2016)
45. Azeemi, I.K., Lewis, M., Tryfonas, T.: Migrating to the cloud: lessons and limitations of 'Traditional'IS success models. In: Conference on Systems Engineering Research (CSER 2013), vol. 16, pp. 737–746 (2013)
46. Chang, V., Walters, R.J., Wills, G.: The development that leads to the cloud computing business framework. Int. J. Inf. Manage. **33**(3), 524–538 (2013)
47. Garg, S.K., Versteeg, S., Buyya, R.: SMICloud: a framework for comparing and ranking cloud services. In: Fourth IEEE International Conference on Utility and Cloud Computing 2011, pp. 210–218 (2011)
48. Ferrer, A.J., Hernández, F., Tordsson, J., Elmroth, E., Ali-eldin, A., Zsigri, C., Sirvent, R., Guitart, J., Badia, R.M., Djemame, K., Ziegler, W., Dimitrakos, T., Nair, S.K., Kousiouris, G., Konstanteli, K., Varvarigou, T., Hudzia, B., Kipp, A., Wesner, S., Corrales, M., Forgó, N., Sharif, T., Sheridan, C., Juan, A., Hernández, F., Tordsson, J., Elmroth, E., Ali-eldin, A., Zsigri, C., Sirvent, R., Guitart, J., Badia, R.M., Djemame, K., Ziegler, W., Dimitrakos, T., Nair, S.K., Kousiouris, G., Konstanteli, K., Varvarigou, T., Hudzia, B., Kipp, A., Wesner, S., Corrales, M., Forgó, N., Sharif, T., Sheridan, C.: OPTIMIS: a holistic approach to cloud service provisioning. Futur. Gener. Comput. Syst. **28**(1), 66–77 (2012)
49. Whaiduzzaman, M., Gani, A., Anuar, N.B., Shiraz, M., Haque, M.N., Haque, I.T.: Cloud service selection using multi-criteria decision analysis. Sci. World J. **2014**, 459375 (2013)
50. Menzel, M., Ranjan, R.: CloudGenius: decision support for web server cloud migration categories and subject descriptors. In: WWW 2012 – Session Web Engineering vol. 2, pp. 979–988 (2012)
51. Kundra, V.: Federal cloud computing strategy (2011)
52. Alonso, J., Orue-Echevarria, L., Escalante, M., Gorronogoitia, J., Presenza, D.: Cloud modernization assessment framework: analyzing the impact of a potential migration to cloud. In: 7th IEEE International Symposium on the Maintenance and Evolution of Service-Oriented and Cloud-Based Systems (MESOCA), pp. 64–73 (2013)
53. Jamshidi, P., Ahmad, A., Pahl, C.: Cloud migration research: a systematic review. IEEE Trans. Cloud Comput. **1**(2), 142–157 (2013)
54. Loebbecke, C., Thomas, B., Ullrich, T.: Assessing cloud readiness: introducing the magic matrices method used by continental AG. In: Nüttgens, M., Gadatsch, A., Kautz, K., Schirmer, I., Blinn, N. (eds.) TDIT 2011. IAICT, vol. 366, pp. 270–281. Springer, Heidelberg (2011). https://doi.org/10.1007/978-3-642-24148-2_18
55. Kauffman, R.J., Ma, D., Yu, M.: A metrics suite for firm-level cloud computing adoption readiness. In: Altmann, J., Vanmechelen, K., Rana, Omer F. (eds.) GECON 2014. LNCS, vol. 8914, pp. 19–35. Springer, Cham (2014). https://doi.org/10.1007/978-3-319-14609-6_2

56. Idris, A.S., Anuar, N., Misron, M.M., Fauzi, F.H.M.: The readiness of cloud computing: a case study in Politeknik Sultan Salahuddin Abdul Aziz Shah, Shah Alam. In: 2014 International Conference on Computational Science and Technology (ICCST 2014), vol. 2014, pp. 1–5 (2015)
57. Olumide, A.A.: Assessment of Cloud Computing Readiness of Financial Institutions in South Africa. University of Cape Town (2014)
58. Kuo, A.M.: Opportunities and challenges of cloud computing to improve health care services. J. Med. Internet Res. **13**(3), e67 (2011)
59. Kaisler, S., Money, W.H., Cohen, S.J.: A decision framework for cloud computing. In: 2012 45th Hawaii International Conference on System Science (HICSS), pp. 1553–1562 (2012)
60. Qian, R., Palvia, P.: Towards an understanding of cloud computing's impact on organizational IT strategy. J. IT Case Appl. Res. **15**(4), 34–54 (2014)
61. Ho, L., Atkins, A.S.: Outsourcing decision-making: a review of strategic frameworks and proposal of a multi-perspective approach. In: Kehal, H.S., Singh, V.P. (eds.) Outsourcing and Offshoring in the 21st Century (IGI 2006), pp. 165–196. Idea Group Inc. (2006)
62. Marketsandmarkets. Healthcare Cloud Computing Market by Application (PACS, EMR, CPOE, RCM, Claims Management), by Deployment (Private, Public), by Service (SaaS, IaaS), by Pricing (Pay as you go), by End-User (Providers, Payers) - Analysis and Global Forecasts to 2020 (2015)
63. Rijnboutt, E., Routsis, D., Venekamp, N., Fulgencio, H., Rezai, M., van der Helm, A.: What challenges have to be faced when using the cloud for e-health services? In: 2013 IEEE 15th International Conference e-Health Networking, Applications and Services (Healthcom 2013), pp. 465–470, October 2013
64. Ratnam, K.A., Dominic, P.D.D.: Adoption of cloud computing to enhance the healthcare services in Malaysia. In: 2014 International Conference on Computational Science, pp. 1–6, June 2014
65. Kuo, M.H., Kushniruk, A., Borycki, E.: Can cloud computing benefit health services? A SWOT analysis. Stud. Health Technol. Inform. **169**, 379–383 (2011)
66. Thuemmler, C., Fan, L., Buchanan, W., Lo, O., Ekonomou, E., Khedim, S.: E-Health: chances and challenges of distributed, service oriented architectures. J. Cyber. Secur. Mobil. **1**, 37–52 (2012)
67. Yoo, S., Kim, S., Kim, T., Baek, R.-M., Suh, C.S., Chung, C.Y., Hwang, H.: Economic analysis of cloud-based desktop virtualization implementation at a hospital. BMC Med. Inform. Decis. Mak. **12**(1), 119 (2012)
68. Yao, Q., Han, X., Ma, X.K., Xue, Y.F., Chen, Y.J., Li, J.S.: Cloud-based hospital information system as a service for grassroots healthcare institutions. J. Med. Syst. **38**(9), 1–8 (2014)
69. COCIR. Advancing healthcare delivery with cloud computing (2012)
70. Caldwell, T.: Two hospitals share a hosted private cloud. Computerweekly (2011)
71. Priyanga, P., Muthukumar, V.P.: Cloud computing for healthcare organisation. Int. J. Multidiscip. Res. Dev. **2**(4), 487–493 (2015)
72. Granados Moreno, P., Joly, Y., Knoppers, B.M.: Public–private partnerships in cloud-computing services in the context of genomic research. Front. Med. **4**, 1–15 (2017)
73. Cianfrocco, M.A., Leschziner, A.E.: Low cost, high performance processing of single particle cryo-electron microscopy data in the cloud. Elife **4**, e06664 (2015)
74. Chen, S.-H., Wen, P.-C., Yang, C.-K.: Business concepts of systemic service innovations in e-Healthcare. Technovation **34**, 1–12 (2014)
75. Hu, Y., Bai, G.: A systematic literature review of cloud computing in EHealth. Heal. Inform. Int. J. **3**(4), 11–20 (2014)

76. Hsieh, J., Hsu, M.-W.: A cloud computing based 12-lead ECG telemedicine service. BMC Med. Inform. Decis. Mak. **12**(1), 77 (2012)
77. Kondoh, H., Teramoto, K., Kawai, T., Mochida, M., Nishimura, M.: Development of the regional EPR and PACS sharing system on the infrastructure of cloud computing technology controlled by patient identifier cross reference manager. Stud. Health Technol. Inform. **192**(1–2), 1073 (2013)
78. Schweitzer, E.J.: Reconciliation of the cloud computing model with US federal electronic health record regulations. J. Am. Med. Inform. Assoc. **19**(2), 161–165 (2011)
79. Rezaeibagha, F., Win, K.T., Susilo, W.: A systematic literature review on security and privacy of electronic health record systems: technical perspectives. Heal. Inf. Manag. J. **44**(3), 23–38 (2015)
80. Lian, J.-W., Yen, D.C., Wang, Y.-T.: An exploratory study to understand the critical factors affecting the decision to adopt cloud computing in Taiwan hospital. Int. J. Inf. Manage. **34**(1), 28–36 (2014)
81. Gao, F., Thiebes, S., Sunyaev, A.: Exploring cloudy collaboration in healthcare: an evaluation framework of cloud computing services for hospitals. In: Proceedings of the Annual Hawaii International Conference on System Sciences, pp. 979–988 (2016)
82. Harfoushi, O., Akhorshaideh, A.H., Aqqad, N., Al Janini, M., Obiedat, R.: Factors affecting the intention of adopting cloud computing in Jordanian hospitals. Commun. Netw. **8**, 88–101 (2016)
83. Lian, J.-W.: Establishing a cloud computing success model for hospitals in Taiwan. Inq. J. Heal. Care Organ. Provision Financ. **54**, 4695801668583 (2017)
84. Griebel, L., Prokosch, H., Köpcke, F., Toddenroth, D., Christoph, J., Leb, I., Engel, I., Sedlmayr, M.: A scoping review of cloud computing in healthcare. BMC Med. Inform. Decis. Mak. **1**(1), 1–16 (2015)
85. IDC: IDC Predicts Saudi ICT Market to Bounce Back in 2017, with NTP Initiatives Spurring Spending to $33.8 Billion (2017). http://www.idc.com/getdoc.jsp?containerId=prCEMA42254117. Accessed 30 Mar 2017
86. Buller, A.: Saudi Arabia enterprise IT trends for 2017. Computerweekly (2016). http://www.computerweekly.com/news/450404528/Saudi-Arabia-enterprise-IT-trends-for-2017. Accessed 30 Mar 2017
87. Oxford Business Group: Positive forecast for data centres and cloud services in Saudi Arabia. Oxford Business Group (2017). https://www.oxfordbusinessgroup.com/analysis/silver-linings-forecast-bright-data-centres-and-cloud-services. Accessed 30 Mar 2017
88. Al-Helayyil, A., Claps, M., Rajan, R., Schaller, O.: Saudi Arabia Vision 2030: Envisioning a Technology-Led Transformation – IDC's Initial View (2016)
89. Vision 2030. Cabinet Approves Kingdom of Saudi Arabia's Vision 2030. Vision 2030 (2016). http://vision2030.gov.sa/en/node/161. Accessed 30 Mar 2017
90. National Transformation Program 2020. National Transformation Program 2020 (2016)
91. Almutairi, N.N., Thuwaini, S.F.: Cloud computing uses for e-government in the middle east region opportunities and challenges. Int. J. Bus. Manag. **10**(4), 60–69 (2015)
92. Yamin, M.: Cloud economy of developing countries. World J. Soc. Sci. **3**(3), 132–142 (2013)
93. Alharbi, S.T.: Users' acceptance of cloud computing in Saudi Arabia. Int. J. Cloud Appl. Comput. **2**(2), 1–11 (2012)
94. Alotaibi, M.B.: Exploring users' attitudes and intentions toward the adoption of cloud computing in Saudi Arabia: an empirical investigation. J. Comput. Sci. **10**(11), 2315–2329 (2014)

95. Alkhater, N., Wills, G., Walters, R., Wills, G., Walters, R.: Factors influencing an organisation's intention to adopt cloud computing in Saudi Arabia. In: 2014 IEEE 6th International Conference on Cloud Computing Technology and Science, pp. 1040–1044 (2014)

96. Alsanea, M., Barth, J.: Factors affecting the adoption of cloud computing in the government sector: a case study of Saudi Arabia. Int. J. Cloud Comput. Serv. Sci. **36**, 1–16 (2014)

97. Alhammadi, A., Stanier, C., Eardley, A.: The determinants of cloud computing adoption in Saudi Arabia. Comput. Sci. Inf. Technol. **2**, 55–67 (2015)

98. Alkhlewi, A., Walters, R., Wills, G.: Success factors for the implementation of a private government cloud in Saudi Arabia. In: Proceedings - 2015 International Conference on Future Internet of Things and Cloud (FiCloud 2015) and 2015 International Conference on Open and Big Data (OBD 2015), pp. 387–390 (2015)

99. AlBar, A.M., Hoque, M.R.: Determinants of Cloud ERP adoption in Saudi Arabia: an empirical study. In: 2015 International Conference on Cloud Computing (ICCC 2015), pp. 1–4 (2015)

100. Tashkandi, A.N., Al-Jabri, I.M.: Cloud computing adoption by higher education institutions in Saudi Arabia: an exploratory study. Cluster Comput. **18**(4), 1527–1537 (2015)

101. Noor, T.H.: Usage and technology acceptance of cloud computing in saudi arabian universities. Int. J. Softw. Eng. Appl. **10**(9), 65–76 (2016)

102. Almutairi, A.A., El Rahman, S.A.: The impact of IBM cloud solutions on students in Saudi Arabia. In: Proceedings of the 2016 4th International Japan-Egypt Conference on Electronic, Communication and Computers (JEC-ECC 2016), pp. 115–118 (2016)

103. Albugmi, A., Walters, R., Wills, G.: A framework for cloud computing adoption by Saudi government overseas agencies. In: 5th International Conference on Future Generation Communication Technologies (FGCT 2016), pp. 1–5 (2016)

104. Mreea, M., Munasinghe, K., Sharma, D.: A strategic decision value model for cloud computing in Saudi Arabia's public sector. In: 2016 IEEE/ACIS 15th International Conference on Computer and Information Science (ICIS), pp. 101–107 (2016)

105. Alharthi, A., Alassafi, M.O., Walters, R.J., Wills, G.B.: An exploratory study for investigating the critical success factors for cloud migration in the Saudi Arabian higher education context. Telemat. Inf. **34**(2), 664–678 (2017)

106. Saudi Genome Project Team. The Saudi Human Genome Program (SHGP). IEEE Pulse, vol. 6, no. 6, pp. 22–26 (2015)

107. Azzedin, F., Mohammed, S., Yazdani, J., Ghaleb, M.: Designing a disease outbreak notification system in Saudi Arabia. In: Second International Conference of Advanced Computer Science and Information Technology (ACSIT 2014), pp. 1–14 (2014)

108. International Futures. Key Development Forecasting- Saudi Arabia. University of Denver (2016). http://www.ifs.du.edu/ifs/frm_CountryProfile.aspx?Country=SA. Accessed 28 June 2016

109. Gale, N.K., Heath, G., Cameron, E., Rashid, S., Redwood, S.: Using the framework method for the analysis of qualitative data in multi-disciplinary health research. BMC Med. Res. Methodol. **13**(1), 117 (2013)

110. Braun, V., Clarke, V.: Using thematic analysis in psychology. Qual. Res. Psychol. **3**(2), 77–101 (2006)

111. Dhar, S.: From outsourcing to cloud computing: evolution of IT services. Manag. Res. Rev. **35**(8), 664–675 (2012)

112. Atkinson, N.L.: Developing a questionnaire to measure perceived attributes of eHealth innovations. Am. J. Health Behav. **31**, 612–621 (2007)

113. Low, C., Chen, Y., Wu, M.: Understanding the determinants of cloud computing adoption. Ind. Manag. Data Syst. **111**, 1006–1023 (2011)
114. HIMSS. 2014 HIMSS Analytics Cloud Survey (2014)
115. Wu, Y.U.N., Cegielski, C.G., Hazen, B.T., Hall, D.J.: Cloud computing in support of supply chain information system infrastructure: understanding when to go to the cloud. J. Supply Chain Manag. Exclus. **49**(3), 25–41 (2013)
116. Haddad, P., Gregory, M., Wickramasinghe, N.: Business value of IT in healthcare. In: Wickramasinghe, N., Al-Hakim, L., Gonzalez, C., Tan, J. (eds.) Lean Thinking for Healthcare. HDIA, pp. 55–81. Springer, New York (2014). https://doi.org/10.1007/978-1-4614-8036-5_5
117. Alshamaila, Y., Papagiannidis, S., Li, F.: Cloud computing adoption by SMEs in the north east of England: A multi-perspective framework. J. Enterp. Inf. Manag. **26**(3), 250–275 (2013)

Author Index

Printed in the United States
By Bookmasters